Detection & Control of Air Pollution

Volume II in MSS' Continuing Series on Air Pollution

Papers by
J. C. Webb, Jean Leroux, Robert S. Braman et al.

MSS Information Corporation
655 Madison Avenue, New York, N.Y. 10021

Library of Congress Cataloging in Publication Data
Main entry under title:

Detection and control of air pollution.

(Air pollution, 2)
1. Air--Pollution--Measurement--Addresses, essays, lectures. 2. Air--Purification--Addresses, essays, lectures. I. Webb, J. C., 1930-
II. Series: Air pollution (New York), 2.
TD890.D47 628.5'3'08 73-12380
ISBN 0-8422-7153-8

COLORADO STATE BOARD OF REGISTRATION FOR PROFESSIONAL ENGINEERS AND LAND SURVEYORS

DPL 132

TABLE OF CONTENTS

CREDITS AND ACKNOWLEDGEMENTS

Braman, Robert S., "Direct Current Discharge Emission Spectra for Detection and Identification of Some Air Pollutant Compounds," *Atmospheric Environment*, 1971, 5:669-675.

Hinkley, E.D.; and P.L. Kelley, "Detection of Air Pollutants with Tunable Diode Lasers," *Science*, 1971, 171:635-639.

Larsen, Ralph I.. "Relating Air Pollutant Effects to Concentration and Control," *Journal of the Air Pollution Control Association*, 1970, 20:214-225.

Lefcoe, Neville M.; and Ion I. Inculet, "Particulates in Domestic Premises. I. Ambient Levels and Central Air Filtration," *Archives of Environmental Health*, 1971, 22:230-238.

Leroux, Jean; and Mazhar Mahmud, "Flexibility of X-ray Emission Spectrography as Adapted to Microanalysis of Air Pollutants," *Journal of the Air Pollution Control Association*, 1970, 20:402-404.

Leroux, J.; and I. Varga, "Average Composition of Particulate Air Pollutants as Disclosed by X-ray Techniques," *International Journal of Environmental Analytical Chemistry*, 1971, 1:131-140.

Minnick, L. John, "Control of Particulate Emissions from Lime Plants — A Survey," *Journal of the Air Pollution Control Association*, 1971, 21:195-200.

Semrau, Konrad T., "Control of Sulfur Oxide Emissions from Primary Copper, Lead and Zinc Smelters — A Critical Review," *Journal of the Air Pollution Control Association*, 1971, 21:185-194.

Slack, A.V.; H.L. Falkenberry; and R.E. Harrington, "Sulfur Oxide Removal from Waste Gases: Lime-Limestone Scrubbing Technology," *Journal of the Air Pollution Control Association*, 1972, 22:159-166.

Stafford, Ronald G.; and Harry J. Ettinger, "Comparison of Filter Media against Liquid and Solid Aerosols," *American Industrial Hygiene Association Journal*, 1971, 32:319-326.

Thaxton, Larry A., "Kish and Fume Control and Collection in a Basic Oxygen Plant," *Journal of the Air Pollution Control Association*, 1970, 20:293-296.

Webb, J.C.; J.C. Kinchen; and J.E. Scarberry, "Aerial Sampling by Helicopter Using a High Volume Sampler," *Journal of the Air Pollution Control Association*, 1970, 20:453-455.

PREFACE

In the United States air pollution reaches toxic thresholds in almost all urban areas. Its cost in terms of life and health is extreme. According to various governmental studies, the annual cost of damage resulting from polluted air is over $13 billion. If, for example, pollution could be cut by 50 percent, newborn babies would have an additional life expectancy of three to five years and deaths would be reduced 4.5 percent. Air pollution is also a factor in the development of diseases ranging from respiratory disorders to cirrhosis of the liver, and polluted air intensifies allergic reactions. Serious long-term research and conscientious political action are needed if these ominous trends are to be reversed.

Volume II in MSS' continuing series on air pollution presents a group of current papers on the detection and control of pollution. General techniques for detection are discussed, including aerial sampling, x-ray techniques, and the use of tunable diode lasers. Special sections are also devoted to the available means of controlling both particulate and gaseous air pollutants.

General Techniques for Detection of Air Pollution

aerial sampling by helicopter using a high volume sampler

J. C. Webb, J. C. Kinchen and **J. E. Scarberry**

Described in this paper is a method of sampling the atmosphere for particulate matter at various elevations above an urban area. Tests were conducted using a high-volume air sampler mounted on a helicopter. Fiber glass filters used to collect the aerosols were analyzed for total particulate, benzene solubles, and specific metals: iron, copper, lead, and zinc. Significant quantities of these aerosols were collected at flight levels of 400 and 800 ft over the city. These results were compared with measurements taken at ground level stations in the same six sampling zones of the City of Fort Worth. The tests demonstrated that a helicopter equipped with a conventional sampler can be effective in evaluating air pollution over an urban area.

Figure 1. Hurricane high volume air sampler mounted on helicopter.

In evaluating the air quality over a city, sampling of the atmosphere is normally conducted at ground stations located throughout the area. These stations consist of a high-volume sampler using a fiber glass type filter and supplied with electric power from 110 volt ac source. Aerosols or particulate concentrations are determined by sampling over a period of time and these values are used to indicate the degree of air pollution.

It is also important to evaluate the concentration of particulates at various levels above the ground, as many of the pollutants will be settling out and becoming a part of the atmosphere we breathe. Aerial sampling has been conducted using fixed wing aircraft and employing the ramjet effect whereby particulates are impinged upon a filter. This type of aircraft is not suitable for an air sampling vehicle over an urban area as there are regulations regarding the flight level which prohibit flying of this type aircraft at levels where most of the air pollution exists.

The City of Fort Worth Health Department has available a helicopter for use in air pollution control activities. FAA regulations permit flight operations of the helicopter as low as 400 ft over the urban area. This aircraft is used by the Department of Public Health in visual surveillance of air pollution over the city. In order to measure the concentrations of pollutants at various levels above the ground, it was decided to equip the helicopter with a high-volume air sampler and conduct tests at two altitudes. This report describes the test operations, the equipment, and presents the data obtained from the use of high-volume sampler in the helicopter.

Test Plan and Engineering

The helicopter used was a 3-passenger Bell, Model 47-G-5. Mounted on this aircraft was a Hurricane high-volume sampler, Model 16003 powered by a McCulloch Mite-E-Lite generator. To install this equipment on the aircraft, it was necessary to obtain approval of the Federal Aviation Authority. The installations of the sampler and generator were accomplished according to the prescribed directions of the FAA.

The generator was mounted on the litter basket located on the left side of the aircraft. Four U bolts were used to secure the generator to a steel plate and to the basket frame. The sampler was mounted, similarly, on the litter basket above the right longeron. A lead plate was added and the two pieces of equipment were positioned so that the weights would be counterbalanced with respect to the aircraft. The sampler was oriented so that the filter would face the direction of flight (Figure 1). A shield was placed over the filter to prevent impact of particulates when the sampler was not in use. The shield was removed at the beginning of each sampling operation. The generator and sampler were mounted as portable equipment. Flight operations were restricted by FAA to sampling missions during the time the equipment was affixed to the craft. Detachment was easily accomplished following a mission and the restriction was removed.

Engineering studies of this particular helicopter have shown that the downwash of the rotor does not disturb the laminar air flow at the position of the sampler provided the speed of the aircraft exceeds 30 mph. The sampling missions were flown at an air speed of 60 mph. During the missions the flow rate, including the ramjet effect, of air passing through the filter was observed to range from 85 to 90 cfm.

To study the distribution of pollutants in the atmosphere over the City, six sampling zones were selected. These were designated as follows:

Zone no.	City area
1	Downtown
2	South of downtown
3	West of downtown
4	North of downtown
5	Northeast of downtown
6	Southwest of downtown

Sampling runs were conducted in each zone at flight levels of 400 and 800 ft above the ground. A crisscross pattern was flown in each zone to obtain representative sampling. During each sampling mission within a zone, the collecting medium used was a Gelman Type A, 8 × 10 in. fiber glass filter. The filters were "conditioned" at 74°F and 35% relative humidity (RH) and weighed before and after each sampling run. The average sampling time for each run was approximately 1.5 hr. The flow rate through the sampler was observed and recorded during the sampling run. Air temperature and sky conditions were noted for each run. In general, throughout the two-week period of testing, the average air temperature at the ground was 93.4°F (USWB); at 400 ft, 88.5°F; at 800 ft, 84.4°F.

Analysis

The concentration of particulates collected by each filter was determined using a Mettler electric balance, Type H6T. A Mettler weighing chamber to accommodate the 8 × 10 in. filters was employed.

The benzene solubles were determined by placing half of the filter in a Soxhlet chamber and extracting with 100 ml benzene for 6 hr. The solution was concentrated and filtered through a Corning fritted glass filter into a pre-weighed aluminum pan. The solvent was evaporated and the organic residues weighed. The samples were analyzed for the following metals: copper, iron, lead, and zinc.

Two-inch strips of the filter were subjected to nitric acid for extraction of the metals. The temperature of the 1:1 nitric acid was maintained at just less than 100°C for 1 hr. The solution was filtered using Whatman No. 42 filter, washed with deionized water and then concentrated to 50 ml. Identification and quantitative determinations of the metals were made using a Beckman atomic absorption system, Model 1301, linked with a DB-E spectrophotometer and recorder.

Results and Conclusions

Data summaries of the sampling conducted at 400 and 800 ft are shown in Tables 1 and 2. Sampling results obtained at selected ground stations during the aerial sampling period are also shown for comparison (Table 3). In review of the average concentrations of particulates it is observed that at 400 ft the amount of aerosol pollutant was about 20% of the ground level average. At 800 ft the particulate concentration was found to be approximately 8% of the ground average. It is significant to note that iron, lead, copper, and zinc were found in measurable quantities at each sampling level.

The tests have shown that a high-volume air sampler can be used effectively in collecting particulate matter carried in the atmosphere over a city. It was also shown that metal compounds of aerosol size and originating from ground sources are transported into the upper air by the atmospheric diffusion process. The tests demonstrated that a conventional high-volume air sampler, mounted on a helicopter can be used to obtain data needed to evaluate the air quality and vital in the control of air pollution over an urban area.

Table I. Aerosol concentrations measured at 800 feet above ground.

(Micrograms/cubic meter of air)

Date	Time	Zone	Particulate	Benzene sol.	Cu	Fe	Pb	Zn
7-7-69	0534–0700	4	33	4.9	0	0	0.688	0.247
7-8-69	0910–1114	1	27	4.5	0	0	0.639	0.202
7-8-69	1400–1536	1	6	2.2	0	0	0.319	0.252
7-9-69	0950–1125	2	11	3.2	0.063	0.243	0.612	Trace
7-9-69	1415–1543	2	<1	0	0	0	0	0
7-11-69	1407–1540	5	<1	0	0	0	0	0
Average			13	2.3	0.01	0.04	0.40	0.12

Table II. Aerosol concentrations measured at 400 feet above ground.

(Micrograms/cubic meter of air)

Date	Time	Zone	Particulate	Benzene sol.	Cu	Fe	Pb	Zn
6-27-69	0543–0731	4	60	14.4	0	1.660	0.958	0.926
6-28-69	0721–0921	1	27	2.6	0.027	0	0.729	0.351
6-30-69	1005–1212	1	16	4.9	0.036	0	0.918	0.184
6-30-69	1428–1555	1	8	2.4	0.027	0	0.715	0.634
7-2-69	0930–1100	4	104	10.2	Trace	0	1.910	0.486
7-2-69	1440–1605	4	43	8.6	0.540	0	1.020	0.549
7-3-69	0915–1045	2	36	2.3	0	0	1.040	0
7-3-69	1845–2020	2	23	7.6	0	0.490	0.612	0.369
7-10-69	1410–1540	4	15	4.8	Trace	0.220	1.870	0.558
7-11-69	0945–1120	5	39	7.3	0.013	0	0.205	0.958
Average			34	5.9	0.06	0.22	0.90	0.46

Table III. Aerosol concentrations measured from ground level sampling (Micrograms/cubic meter of air)

Date	Time	Zone	Particulate	Benzene sol.	Cu	Fe	Pb	Zn
[a]6-11-69	0758–0802	3	122	1.9	0.946	0.737	0.431	0.013
[a]6-13-69	0800–0800	5	62	4.6	0.146	0.519	2.78	0.556
6-30-69	1020–1200	1	129	2.9	0.026	1.29	1.62	0.251
6-30-69	1400–1530	1	82	1.6	0	10.03	0	0.324
7-2-69	0920–1700	1	63	2.2	0.189	0.496	0.496	0.044
7-3-69	0915–1123	1	63	3.4	Trace	Trace	0.426	0.019
[a]7-9-69	1035–855	1	937	20.1	0.002	9.98	8.9	0.647
7-10-69	0750–0920	b	30	4.8	0	2.08	Trace	0.720
7-10-69	1105–1229	b	94	11.8	Trace	0	17.01	0.540
Average			175	5.9	0.15	2.79	3.50	0.81

[a] 24-hr sample.
[b] Sample taken in non-urban area, 20 miles from downtown area.

Bibliography

1. Drees, J. M. and Hendal, W. P., "Air flow patterns in neighborhood of rotor," *Air Craft Engr.* (1951).
2. Brotherhood, P. and Stewart, W., "An Experimental Investigation of The Flow Through a Helicopter in Forward Flight," *Rep. Aeron. Res. Council (London)*, No. 2734.
3. Clayton, G. D., *Determination of Atmospheric Contaminates*, New York, Amer. Gas Assoc., 1953.
4. Gartrell, F. E. and Carpenter, S. B., "Aerial sampling by helicopter: A method for study of diffusion patterns," *J. Meteorol.* **12**, 215 (June 1955).
5. Gassman, G. J. and Pittenger, E. W., "Upper Atmosphere Research Using Aircraft," AF CRL-66-292, Special Reports, No. 44, Air Force Cambridge Research Laboratories, Bedford, Mass., 1966.
6. Lee, R. F., Patterson, R. K., and Wagman, J., "Particle-size distribution of metals components in urban air," *Environ. Sci. Technol.* **2**, 288 (Apr 1968).
7. Stanley, T. W., Meeker, J. E., and Morgan, M. J., "Extraction of organics from airborne particulates," *Environ. Sci. Technol.* **1**, 927 (Nov. 1967).
8. *Major Repairs and Alterations*, FAA Form 337, Part 43 U. S. Federal Aviation Regulations.

Flexibility of X-Ray Emission Spectrography as Adapted to Microanalysis of Air Pollutants

Jean Leroux and Mazhar Mahmud

In a preceding paper "Preparation of thin dust coatings for their analysis by x-ray emission and diffraction,"[1] it was stressed that both techniques, as applied to industrial and urban atmospheric dusts, are ideal for the following reasons: they require little ancillary equipment besides the x-ray instrumentation while their degree of accuracy, sensitivity and speed of analysis permit their use with only occasional checks by other popular but time-consuming techniques.

These advantages have been put forward in other applications involving analysis of thin film samples as reviewed by Campbell[2] and have been exploited earlier with fair success by R. C. Hirt, et al.,[3] in the analysis of dusty atmospheres. The main difficulty of these last applications lies in the lack of an easy solution for preparing homogeneous standards of reference and for discarding uncertainties in the determination of elements due to matrix effects of a given sample.

The present paper offers a technique which eliminates these disadvantages, being itself a specific application of the investigation reported above[1] and of "X-ray quantitative analysis by an emission-transmission method."[4] As more detailed information pertaining to its theoretical implications, standard sample preparation, and sampling conditions can be found in the above-mentioned papers, these aspects will hereafter be referred to only when needed

New Experimental Expression

In the original paper,[4] the concentration of a given element i in a sample is expressed as

$$C_i = \frac{I_i \rho_i}{K I_0 W} \phi(T) \qquad (1)$$

where I_i and ρ_i are the intensity of the radiation emitted by and density of i respectively, K an excitation constant, I_0 the intensity of the primary beam, W the weight of the sample, and $\phi(T)$ a function of the transmittance T which refers to a transmission measurement.

Since $C_i = W_i/W$, then from Equation 1, one obtains

$$W_i = \frac{I_i \rho_i}{K I_0} \phi(T) \qquad (2)$$

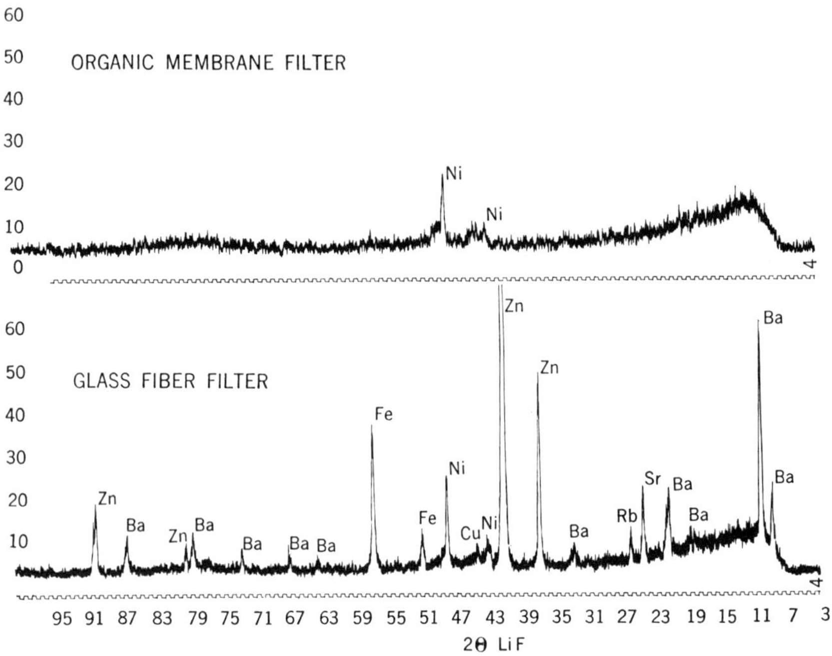

Figure 1. X-ray emission tracings showing chemical composition of organic membrane and glass fiber filters.

Equation 2 as applied to a standard sample can be written

$$W_y = \frac{I_y \rho_i}{K I_0} \phi(T_y) \qquad (3)$$

and as applied to an unknown sample

$$W_x = \frac{I_x \rho_i}{K I_0} \phi(T_x) \qquad (4)$$

Then from (3) and (4)

$$W_x = W_y \cdot \frac{I_x}{I_y} \cdot \frac{\phi(T_x)}{\phi(T_y)} \qquad (5)$$

Since W_y and $\phi(T_y)$ are known and constant for definite and fixed values of voltage and amperage of the x-ray target one has the final expression

$$W_x = \frac{C I_x}{I_y} \cdot \phi(T_x) \qquad (6)$$

In this experimental equation W_x can be considered as a linear function of I_x/I_y for a value of $\phi(T_x)$ close to unity.

In Table I are shown calculated values of T and corresponding values $\phi(T)$ and $1/\phi(T)$ indicating the error percent made in the weight determination when transmission measurement is neglected.

Fortunately, the transmittance value calculated for the average air pollution samples is rarely below 0.80 for radiations emitted by elements higher than manganese in the Periodic table. Referring to Table I, the error made in weight by neglecting the transmittance of 0.80 for a sample containing 20 μg of a given element would be 10%, that is 2 μg and might be considered as negligible when values are expressed in micrograms per cubic meter of air sampled.

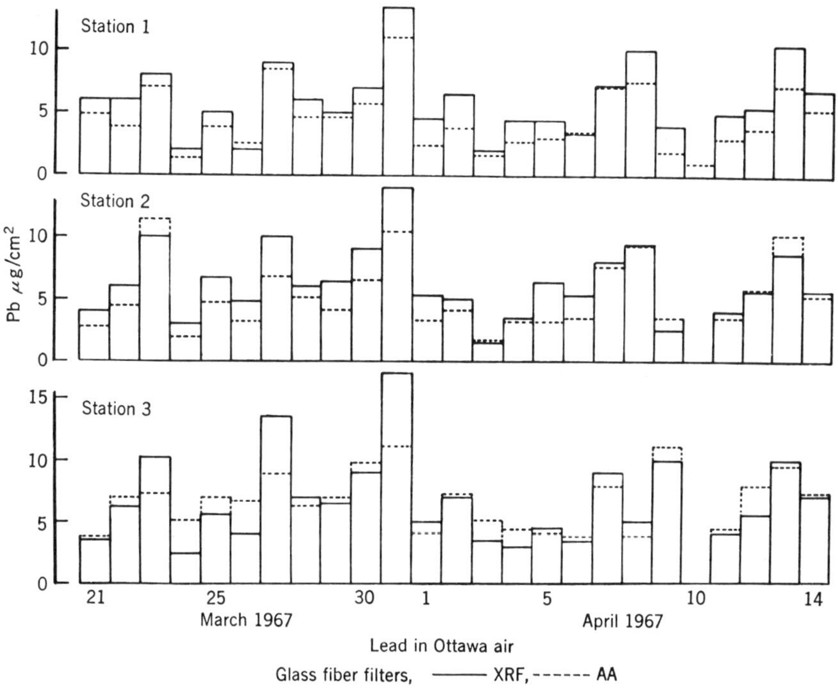

Figure 2. Comparison analysis obtained by x-ray emission (——) and atomic absorption (----).

Choice of Filtering Material

For analysis of trace elements, the chemical composition of filters used in air pollution surveys is of paramount importance. It should be free of any element presumably present in the dust sample itself. The latter does not usually exceed more than 10 mg/cm^2 and thus is unable to absorb the interfering x-ray lines emitted by the supporting filter. Figure 1 shows the x-ray emission tracings of the two most commonly used filters; i.e., organic membrane and glass fiber filters. The absence in these two tracings of the tungsten characteristic lines usually scattered by material of such low atomic numbers is due to their removal from the incident beam by nickel foils according to a technique published by this laboratory;[5] the two remaining faint lines observed are those scattered by the nickel foils. These tracings indicate that the glass fiber filters are definitely not suitable for multielemental analysis by x-ray emission because of their content of Ba, Sr, Rb, Zn, Ni, Fe, Ca, etc. However, they have been found to be fairly good for analysis of lead in the air pollution survey described below.

Applications

Lead Analysis on Glass Fiber Filters

In the early stage of this investigation and with a view to comparing the sensitivity and accuracy of this new technique with those of atomic absorption, hundreds of samples have been tested for lead by both techniques.

All samples were collected in Ottawa

at three experimental stations approximately 2000 ft apart in a triangular pattern. This survey was made continuously in the first 5 months of 1967 at the rate of one sample per day at each station. Each sample was collected by means of a high volume sampler on Gelman glass fiber filters 8 in. × 10 in. in size. Lead analysis of about twenty discs, 1¼-in. diameter, cut at random over the total area, demonstrated that the dust coating was very uniform. Therefore the average of the results of lead analysis obtained with two discs cut from each sample was found sufficient.

Values of lead expressed in $\mu g/cm^2$ for the three stations over a period of 25 consecutive days are shown in Figure 2 in dark lines for x-ray and dotted lines for atomic absorption methods. The mean deviation in $\mu g/cm^2$ calculated between results of both techniques over a period of 60 consecutive days of sampling is 1.25, 1.15, and 1.17 for stations 1, 2, and 3, respectively. These values are in the order of 0.26 when expressed in $\mu g/m^3$. As observed in Figure 2, the slight deviation in the lead content measured by both techniques for that period barely upsets the agreement in the day-to-day fluctuations.

Table I. Calculated values of T, $\phi(T)$, $1/\phi(T)$ and error % made in the weight by neglecting the transmission measurement.

T	$\phi(T)$	$1/\phi(T)$	Error %
1.00	1.00	1.00	0
0.98	1.01	0.99	1
0.96	1.02	0.98	2
0.95	1.026	0.97	3
0.90	1.054	0.95	5
0.85	1.083	0.92	8
0.80	1.116	0.90	10
0.75	1.151	0.87	13
0.70	1.189	0.84	16
0.65	1.231	0.81	19
0.60	1.277	0.78	22
0.40	1.527	0.65	35
0.30	1.720	0.58	42
0.20	2.012	0.50	50
0.10	2.558	0.39	61

The transmission of the lead radiation used (1.175 A) being very close to 100%, as checked occasionally from a few samples in each series, all series were analyzed by simply comparing I_x and I_y on the instrument, the lead content per cm^2 being derived from Equation 6 which in this case becomes:

$$W_x = 47.8\frac{I_x}{I_y} \qquad (7)$$

where $C = [W_y/\phi(T_y)] = (59.8\ \mu g\ Pb/1.25)$. The standard of reference was a pressed boric acid pellet containing 0.1% of PbO, 1¼ in. in diameter and weighing 0.510 g.

Analysis of Lead, Zinc, Iron, Manganese, Calcium, and Sulfur on Organic Membrane Filters

As discussed in reference (1), sampling of dusty atmospheres through filters 37 mm in diameter is ideally achieved with disposable plastic monitors such as those available from Millipore. The filter is located between two cylindrical chambers, the top one being used as the monitor lid after sampling. However, the lid is usually replaced by a retaining plastic ring in order for the dust to be uniformly distributed over the total area of the filter. Best results are obtained in these conditions with an air flow rate of approximately 2 l/min.

As an example of multi-elemental analysis performed with these filters, we have selected five typical analyses made on a large scale survey in Montreal at stations located in heavy traffic zones.

Since this x-ray method gives directly the amount of a given element, weighing of the filter before and after sampling is not essential and its weight can be expressed in $\mu g/m^3$ of air sampled. However, the five samples have been weighed to illustrate the relative importance of the matrix effect according to their weight per square centimeter of dust deposit. Table II lists these weights expressed in $\mu g/cm^2$ with those determined for all elements; these values have been converted to $\mu g/m^3$

Table II. Analysis of lead, zinc, iron, manganese, calcium, and sulfur in air pollution samples collected in the Montreal area in 1967 and 1968.

Sample no.	Micrograms/cm^2													Micrograms/m^3						
	Total dust	Pb	T[a]	Zn	T	Fe	T	Mn	T	Ca	T	S	T	Total dust	Pb	Zn	Fe	Mn	Ca	S
1	850	2.0	0.99	0.78	0.94	2.47	0.92	0.25	0.79	8.1	0.85	46	0.43	99	0.28	0.11	0.35	0.30	1.11	6.44
2	1000	2.5	0.99	1.10	0.94	1.87	0.92	0.25	0.79	6.3	0.85	42	0.43	166	0.35	0.15	0.26	0.03	0.87	5.9
3	1060	2.6	0.99	1.48	0.97	2.66	0.94	0.31	0.84	6.3	0.85	56	0.40	123	0.36	0.20	0.37	0.04	0.87	7.75
4	1100	2.0	0.99	0.58	0.98	2.16	0.97	0.19	0.83	15.	0.83	26	0.47	167	0.28	0.08	0.30	0.03	2.06	3.64
5	1540	7.7	0.99	29.3	0.93	6.00	0.92	0.31	0.75	32.	0.73	184	0.27	234	1.07	4.03	0.83	0.04	4.25	25.6

Standard sample	Tested element	$C = [W_y/\phi(T_y)]$	Detector	Crystal analyzer
Pellet of boric Acid				
0.1% PbO	Pb	59.8/1.25	Scintillation counter	Lithium fluoride
0.1% ZnO	Zn	51/1.58	Scintillation counter	Lithium fluoride
0.1% $MnCO_3$	Mn	31.2/3.31	Scintillation counter	Lithium fluoride
Dust coating of pure $CaCO_3$	Ca	163/1.1	Flow counter	EDTT (in vacuum)
Dust coating of pure Sulfur	S	440/1.276	Flow counter	EDTT (in vacuum)

[a] Transmittance derived from transmission measurements. Its function $\phi(T)$ is used as a matrix correction factor of the intensity measurement.
A tungsten target operated at 50 kv, 20 mA was used in all these analyses.

on the right hand side of the table. The transmittance value corresponding to that calculated for each element and each sample is also indicated besides each weight measurement.

Under this table are indicated the experimental conditions of analysis, the physical characteristics of standard samples, and the *C* value to be used in the experimental equation.

These results suggest that the average air pollution samples weighing less than 1.5 mg/cm^2 can be analyzed practically without matrix correction for all its elements higher than manganese in the Periodic table.

Conclusion

This microtechnique first reported verbally at a meeting in 1968,[6] has since been applied to a large number of environmental air pollutant problems yielding the following conclusions.

Due to the small area of coating irradiated by x-ray (usually 1.5 cm^2) the size of the filter in the monitor can be as small as 7 cm^2. Under these conditions, shorter intervals of sampling are obtained as compared to the 24 hr sampling time required with conventional hi-volume sampler using 8 $\times$ 10 in. glass fiber filters.

The method being non-destructive, the sample is analyzed directly as received, thereby decreasing the number of possible experimental errors accumulated through a sequence of manipulations usually involved in titration, acid digestion, extraction, etc. On the other hand, its accuracy in the range of sensitivity required for an average air pollution sample has been proved at least as good as that of atomic absorption in the lead survey described in this paper. In terms of x-ray analysis, major delay is in the sampling time itself, since approximately 5 min is required for the analysis of a single element after receipt of the sample, as compared to several hours for other conventional spectrographic techniques.

However, while the method described in this paper is adaptable to the more recent multi-channels and fully automatic x-ray emission spectrometers, it is believed that the addition of the latter to the basic equipment of an air pollution network would add still more efficiency to the conduct of a large-scale and long-term atmospheric survey.

References

1. Leroux, J., *Staub Reinhaltung der Luft*, **29,** 4 (1969).
2. Campbell, W. J. and Brown, J. D., *Anal. Chem.* **40,** 346R (1968).
3. Hirt, R. C., Doughman, W. R., and Gisclard, J. B., *Anal. Chem.* **28,** 1649 (1956).
4. Leroux, J. and Mahmud, M., *Anal. Chem.* **38,** 76 (1966).
5. Leroux, J. and Mahmud, M., *Can. Spec.* **13,** 4 (1967).
6. Leroux, J., Analysis of Microgram Quantities of Industrial and Air Pollutants Samples by X-ray Diffraction and Emission. 15th Spectroscopy Symposium of Canada, October 21–23, 1968, Toronto.

Average Composition of Particulate Air Pollutants as Disclosed by X-ray Techniques

J. LEROUX and I. VARGA

Analysis of airborne particulates published for the last 18 years have been assembled thus disclosing a definite range of concentrations for many elements which could be statistically determined, at least when related to the atmosphere of many industrial cities of the northern hemisphere. This view has been reasonably well confirmed by X-ray absorption curves which were computed, plotted for hypothetical samples and compared with those obtained experimentally with a large amount of actual samples. The absorption curve obtained for the average hypothetical sample thus reconstituted suggests two practical applications. The technique might be used as a fingerprint for tracing abnormal but unsuspected composition of a given sample and for the weighing of thin dust deposits with less probable error than with the usual gravimetric methods. The adaptation of the technique to these possible applications is discussed.

One important phase involved in the control of air pollution is the study of the composition of airborne particulates in specific areas. At first sight, such a study may appear as an overwhelming task, especially if one considers all the possible natural and man-produced sources of that material, as well as the number of physical and meteorological factors contributing to their diffusion in the atmosphere. The problem considered in this light may render the selection of analytical techniques and instrumentation very difficult due to the unpredictable presence of compounds in a given sample. However, the complexity of the problem could be considerably reduced if some knowledge of the range of concentration and nature of the main elements usually found in the atmosphere could be attained with a reasonable degree of confidence.

The possible existence of common denominators, at least between the crystalline forms of some compounds identified in various cities, has been previously reported in 1954.[1] Dust samples collected at that time in Windsor, Ottawa, Paris, Marseille, Rome, Venice, Milan, and Madrid were analyzed by X-ray diffraction and showed the constant and common presence of quartz, gypsum, anhydrous calcium sulfate, calcite, feldspar, magnetite and hematite. Similar analytical results have since been obtained for samples collected in Tokyo, Detroit, New York, and Montreal, including one from Resolute Bay—400 miles from the North Pole.

Although one can find in the literature some reports evaluating the constant recurrence of compounds commonly found in air samples, these are generally limited to specific groups such as the organic fraction[2] or metallic compounds[3] emitted in the air by industrial processes. The main purpose of the investigation described herein was to simulate a hypothetical air pollution sample with an elemental composition comparable to that of samples selected at random from significantly different geographical locations.

THEORETICAL APPROACH

At the outset of this investigation, data obtained from our previous experience were assembled with those from reports published by national and international organizations concerned with air pollution. After an extensive comparison of these data, it seemed obvious that a definite range of concentrations of many elements could be statistically determined at least when related to the atmosphere of many industrial cities located in the northern hemisphere.

In order to test the validity of this view, a hypothetical model of an air pollution sample with a composition approximating that of a so-called average sample, was first established according to the above-mentioned references. By keeping the nature of this sample constant and changing the relative concentration of its components, one obtains substances more or less permeable to a monochromatic X-radiation. The technique of X-ray absorptiometry, which is based on that property of matter to absorb differentially monochromatic X-radiations according to their wavelength and to the atomic number of its constituents, was selected in this investigation for the following reasons. The capacity for a substance to absorb a given monochromatic X-radiation can be predicted[4] with a reasonable degree of accuracy within a practical range of wavelengths. Under these conditions, an infinite number of combinations of mixtures can be computed for their absorption as a function of the wavelength, thereby making unnecessary such determination on an equivalent number of experimental synthetic samples. Because the basic laws governing techniques such as X-ray absorptiometry or X-ray transmission measurements have been well described in current literature,[4–7] the expressions necessary only to the understanding of this paper will be recalled in the following.

The ratio of the energy dissipated by a monochromatic beam of X-rays through matter is usually expressed as

$$I/I_0 = -\mu \rho x \tag{1}$$

where I_0 is the incident beam, I the transmitted beam, and μ, ρ, and x are respectively the mass absorption coefficient/$cm^2\ gm^{-1}$ of the absorber, its density, and its thickness.

The empirical expression relating μ to λ (wavelength of the X-ray beam) has the form

$$\mu = C\lambda^n \tag{2}$$

where C and n are constants having definite values[4] for elements of specific atomic number Z and between definite ranges of wavelengths called "absorption edges" of any specific element. An absorption edge is characterized by a sharp drop of absorption in the substance for a radiation with a slightly higher wavelength than the absorption edge value.

In the case of a mixture, the mass absorption coefficient (MAC) is

$$\mu_m = \mu_1 x_1 + \mu_2 x_2 + \cdots + \mu_n x_n = \sum \mu_i x_i \tag{3}$$

where μ_i is derived from Eq. (2) and where x_i corresponds to the weight fraction of the element in the mixture for a given wavelength.

On the other hand, as was the case in the last stage of this investigation, different values of μ_m in Eq. (3) can be obtained for the same wavelength by separating all elements into four distinct groups, e.g. A, B, C and D. Keeping the weight fraction of elements within each group constant and designating by C the group concentration, one has

$$\mu_m = \mu_A C_A + \mu_B C_B + \mu_C C_C + \mu_C C_D \tag{4}$$

The general procedure described hereafter was based on the above theoretical considerations.

PROCEDURE

The sequence followed in this entire investigation was, by necessity, of an iterative nature. A series of values was computed, plotted and compared to a curve traced through the average experimental MAC of actual samples expressed as a function of wavelength. This process was repeated until a reasonable agreement was obtained between experimental and calculated curves.

Evaluation of the first hypothetical mixture

Using the relevant data which had been gathered previously, the first approach was to eliminate any hypothetical composition of nonrealistic occurrence such as, reasoning *ab absurdo*, one made of 5% diamond dust and 95% mercury. In fact, it appeared quite distinctly that a total of 25 elements could be retained, many others being reported as traces only with no possible bearing effect on subsequent calculations.

Another significant observation was that when these elements were arranged in ascending order of atomic numbers, they could be divided into four groups, with the relative concentrations of the groups being in a descending order. Elements contained in Groups A, B, C and D were identical to those listed in Table I, although this table does not indicate their originally selected weight fraction. Group A includes hydrogen, carbon and oxygen which are assumed to belong to material such as hydrocarbons, oil, tar, water and radicals or oxides of elements found in other groups. Group B contains elements associated with compounds such as quartz, silicates, feldspar, sulfates, vitreous slags, etc., disseminated in the atmosphere by natural as well as industrial processes. Group C was assumed to be associated with pollutants produced mainly in heavy industrial areas where iron was considered by far the most important. The five elements listed in Group D had the same total concentration of 1.5% as shown in Table I in the first tested mixture with lead being the dominating one.

Calculation of the mass absorption coefficients of the first hypothetical mixture

The calculations were made using a program written in Fortran IV language which was submitted to an I.B.M. 360 Model Computer. The program contained instructions for calculating, by means of Equations 2 and 3, fluctuations of μ_m as a function of the nine different wavelengths, as applied to the 25 elements respectively. A total of 1200 MAC values were calculated, representing some 150 hypothetical samples. Only those values which fell within the shaded area of Figure 1 were retained for further recalculations. The upper and lower boundaries of the shaded area correspond respectively to the minimum and maximum MAC values determined from transmission measurements on actual samples; the mean values are designated by circles.

TABLE I

Comparison in the variation of μ as a function of λ for hypothetical Mixture No. I and for a series of dust samples collected in different cities. $\mu_g = \mu_i C_i$ within a given group and μ_g/μ_m is its relative contribution to the total μ of the mixture.

(A) Group	Elements and their concentration % in Mixture I	λ	Mo Kα 0.71	Sr Kα 0.88	Pb L. 1.175	Zn Kα 1.435	Cu Kα 1.54	Fe Kα 1.937	Cr Kα 2.291	Ti Kα 2.748	Ca Kα 3.359
			μ_g and μ_g/μ_m								
A	H 5, C 30, O 27	Gr % 62	0.757 9.2%	1.206 9.6%	2.466 9.4%	4.17 9.4%	5.074 9.6%	9.483 13%	15.086 13.1%	25.06 13.5%	44.104 17.9%
B	Na 1.2, Mg 1.2, Al 1.2, Si 13.6, S 6, Cl 0.2, K 1.8, Ca 4.8	30	2.89 35%	5.3 42%	12 46%	21.12 48%	25.9 49%	49.4 68%	79.4 69%	133 72%	169 69%
C	Ti 0.5, V 0.01, Cr 0.01, Mn 0.01, Fe 5.4, Co 0.02, Ni 0.02, Cu 0.03, Zn 0.5	6.5	3.1 38%	4.46 35.4%	10.1 38%	15.98 36%	18.4 35%	6.72 9%	10.46 9%	12.17 7%	20.78 8%
D	As 0.05, Br 0.1, Sr 0.05, Sn 0.5, Pb 0.8	1.5	1.46 18%	1.6 13%	1.76 7%	2.95 7%	3.55 7%	6.52 9%	9.83 8%	15.7 8%	12.44 5%
		μ_m	8.2	12.6	26.326	44.22	52.92	72.123	114.78	185.9	246.3
(B)	Experimental values obtained with samples from: Detroit, New York, Ottawa, Paris and Tokyo.	Mean	6.6	11.4	24.65	37.67	51.2	70.6	111	183	218
		Minimum	6.15	10.13	24.4	32.	38.	54.1	84.4	145.5	200.4
		Maximum	6.8	11.84	27.0	40.8	65.5	83.6	140.0	207.0	235.0

Calculations of the final hypothetical mixture I

In the first attempts at obtaining a theoretical curve matching the curve traced between the average experimental points, it was found that none of them was entirely satisfactory. While most points closely followed the experimental curve for a large range of wavelengths, sudden deviations of the MAC values corresponding to definite wavelengths were deemed unacceptable. These discrepancies are due to the absorption edge effects

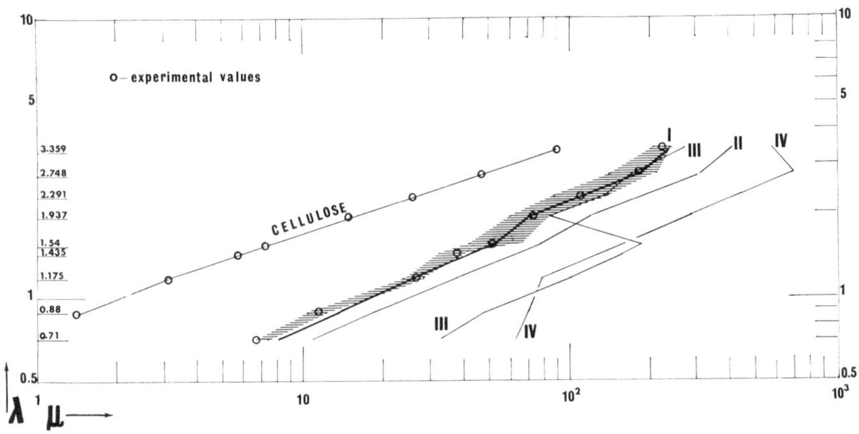

FIGURE 1 Mass absorption coefficient versus wavelength of the four hypothetical Mixtures I, II, III and IV. The upper and lower boundaries of the shaded area correspond respectively to the minimum and maximum experimental values.

explained previously. Since each element is characterized by its own absorption edges it can be expected, with an element abnormally high in concentration, that the calculated MAC value of the total sample will automatically be far from the average for a wavelength close to an absorption edge. It was then realized that slight variations in the concentrations of Si, K and Ca in Group B; Ti, Fe and Zn in Group C; and of Pb in Group D were largely responsible for the trend of these deviations. Therefore, these were used as key elements for determining the final weight fraction of all other elements included in their respective groups.

Finally, by means of Eq. (4), the definitive curve (bold line on Figure 1) relating μ_m to λ for the hypothetical Mixture I was traced. In Table I under each wavelength value, are listed the corresponding MAC of each group, μ_g, and its relative contribution, μ_g/μ_m (expressed in percent), to the total μ_m of the mixture.

Comparison of mixture I with other extreme hypothetical mixtures

Figure 1 illustrates to what extent different relative concentrations in hypothetical samples can modify the curve of mass absorption coefficient versus wavelength. As shown in Table II, the maximum concentration of 62% of Group A in Mixture I has been transferred to Groups B, C and D in the Mixtures II, III and IV, respectively. Results obtained by recalculation using Equation 4 with the concentrations thus modified yielded the three corresponding curves II, III and IV traced on Figure 1. In addition, the extreme case for a light mixture which would be made only of C, H and O has been calculated and measured experimentally with a pure cellulose sample, resulting in the curve designated Cellulose in Figure 1. This specific test was made purposely to demonstrate the predictability of the X-ray transmission measurements when applied to a sample made of low atomic elements.

TABLE II

Percentage of each group used for calculating absorption of hypothetical Mixtures I, II, III and IV, the relative amount of elements within each group being kept equal to that used in Mixture I.

Group	Amount % in mixtures I	II	III	IV
A	62	30	30	30
B	30	62	6.5	6.5
C	6.5	6.5	62	1.5
D	1.5	1.5	1.5	62
Total	100	100	100	100

EXPERIMENTAL

Most tested samples originated from the deposits of particulate material collected from screen filters commonly used in the ventilation systems of large buildings located in the afore-mentioned cities. Notwithstanding the year of their collection (between 1953 and 1970), the season, the continent, the country or any particular site within a city where they were collected, all samples, without exception, presented a common X-ray diffraction pattern. The main differences were in colour, grayish to black, and consistency mainly attributed to variations in the amount of carbon black, hydrocarbons and organic fibrous material. Removal of the benzene soluble material followed by the pulverization of the remaining solid residue in liquid nitrogen, converted all samplesin to a finely divided material ready for transmission measurements.

Transmission measurement

The samples were prepared for transmission measurements by coating them on membrane filters according to a technique developed in this laboratory.[8] Each sample prepared by this technique was very uniform in thickness and weight, and was measured for its MAC by usual X-ray transmission measurements,[6,7,9,10] employing a Philips High Vacuum Spectrograph. Four membrane filters were coated with each air pollution sample originating from the cities indicated in Table I. The monochromatic radiations used were those emitted by samples containing molybdenum, strontium, lead, zinc, copper, iron, chromium and calcium: the respective wavelengths of which are listed in Table I in ascending order. The experimental MAC values (minimum, average, maximum) obtained from these transmission measurements and application of Eq. (1) are listed in Table I (B) directly below the calculated values for Mixture I.

X-ray diffraction and fluorescence analysis

Of the twenty-five elements chosen to satisfy hypothetical Mixture I, Si, Fe, Pb, S, Ca and K are among the most important in terms of concentration. It seemed interesting to measure, in the experimental samples, the actual concentration of quartz (SiO_2) by X-ray diffraction,[6,7] and of the other elements by X-ray fluorescence.[9,11] The average concentration found for quartz was 10.4%; for Fe, 7.1%; Pb, 1.14%; S, 5.8%; and Ca, 2.3%. In the estimated 13.6% total silicon indicated in Table I, approximately 5% must belong to quartz, the remaining 8.74% being associated with silicates and vitreous material. The average deviation between estimated and corresponding experimental percentages of other elements is 1.25%.

Combustible material

The average benzene soluble material was found to be in a proportion of 6.0% while an average of 41% was lost by incineration at 600°C. According to Table I, Group A should contain elements associated mainly with the combustible material, that is, hydrogen, carbon and oxygen. Its estimated concentration is 62%. The oxygen weight fraction is evaluated at 27%; approximately 15.5% pertains to SiO_2 (total silica), the remaining 11.5% being distributed among organic and inorganic compounds.

DISCUSSION

Examination of Figure 1 yields evidence that airborne particulates found in industrial cities have comparable average elemental composition and distribution which can be revealed by X-ray absorptiometry.

This is substantiated by the absorption curves of hypothetical Mixtures I, II, III and IV. The only difference between Mixture I and II is the interchange of concentrations of Groups A and B, which results in an increase of the absorption from a factor of 1.3 for short wavelengths to a factor of about 1.8 for the longer wavelengths. In Mixture III, which has a maximum concentration for Group C, the absorption increases four times for the short radiations, drops to 1.14 for radiations larger than 1.54 (CuKα), becomes less for the titanium radiation (2.748), then increases again by a factor of 1.13 the calcium radiation. The first drop is due mainly to the higher concentration in Group C of iron, which is a very low absorbing material for its own radiation; other variations in the range of the titanium and calcium radiations are due to the increase of titanium in Group C and to the decrease of calcium in Group B, respectively. Similarly the curve of Mixture IV, where lead is predominating in Group C, is definitely more absorbing than other mixtures at all wavelengths. In this case, the rate of change of both variables μ and λ, as applied to pure lead can be well perceived with the general trend of the curve.

The entire procedure described in this paper is primarily related to statistical aspects rather than to a specific method of analysis of air pollutants. However, the results obtained suggest that beyond academic interest it might, in fact, yield genuine practical applications deserving further study. Two such applications have already been approached and will be discussed in more detail in subsequent papers. The first approach concerns the application of the absorptiometry technique to air pollutant samples as a fingerprint for determining abnormal but unsuspected composition of a given sample. In fact, periodic comparison of a sample absorption curve with that of Mixture I might disclose an abnormally high specific source of air pollution in a particular zone.

The second application would permit weighing thin dust deposits on cellulosic membrane filters, provided the composition of the sample is of the average type as discussed herein. Since the MAC value of such a sample shows a deviation of $\pm 7\%$ in the range of soft radiations (2.748 to 3.359), its weight/cm^2 can be calculated easily by transmission measurements. Factors such as electrostatic charges, humidity and filter contamination, which might decrease the accuracy of gravimetric methods by as much as 300%, have practically no effect on X-ray measurements. Therefore, this X-ray technique might conceivably replace the former in the weight determination of microgram quantities of airborne particulates which have the physical characteristics of the hypothetical sample described in this investigation.

References

1. J. Leroux et J. J. Trillat, *J. Rech, C.N.R.S.*, **27**, 389–391, (1954).
2. L. Dubois, A. Zdrojewski, P. Jennavar and J. L. Monkman, *Atm. Environment* **24**, 199–207, (1970).
3. E. C. Tabor and W. V. Warren, *Am. Ind. Health* **17**, 145–151, (1958).
4. J. Leroux, *Adv. X-Ray Anal.* **5**, 153–160, (1961).
5. H. A. Liebhafsky, H. G. Pfeiffer, E. H. Winslow and P. D. Zemany, *X-ray Absorption and Emission in Anal. Chem.* (John Wiley and Sons), 314–317, (1960).
6. J. Leroux, D. H. Lennox and K. Kay, *Anal. Chem.* **25**, 741–743, (1953).
7. J. Leroux, *Norelco Reporter* **4**, 107, (1957).
8. J. Leroux, *Staub—Reinhaltung der Luft* **4**, (1969)—English Version.
9. J. Leroux and M. Mahmud, *Anal. Chem.* **38**, 76, (1966).
10. J. Leroux, M. Mahmud, A. B. C. Davey, *Can. Spec.* **12**, 5, (1967).
11. J. Leroux and M. Mahmud, *Air Pollution Control Association* J. 20, **6**, 402–404, (1970).

DIRECT CURRENT DISCHARGE EMISSION SPECTRA FOR DETECTION AND IDENTIFICATION OF SOME AIR POLLUTANT COMPOUNDS*

ROBERT S. BRAMAN

Abstract—A four-channel d.c. discharge emission type detector was used in a study of the emission response of nitrogen oxides, CO, CO_2, COS, SO_2 and selected organic compounds. Limits of detection for the inorganic gases were determined and found to be in the 10^{-9} to 10^{-11} l. range. Identification of compounds by a normalized relative response ratio technique was studied and found to be feasible. The structure and chain length of a homologous series of compounds influences the comparative response obtained at the emission wave lengths of the C_2, CH, CN bands and H line selected for study. The elemental composition of the inorganic gases was reflected in the relative responses obtained.

INTRODUCTION

THE PURPOSE of this study has been to investigate the use of d.c. discharge emission-type detectors in air pollutant analysis and detection in a gas chromatographic system. The characteristic lower limits of detection (10^{-10} to 10^{-16} g s^{-1}) and capability for qualitative identification of sample components of the d.c. discharge emission type detector as developed by MCCORMACK *et al.*, 1965 and BRAMAN (1968) recommended the study.

When a d.c. discharge is used in helium carrier gas, as in the present case, a great variety of both atomic line and diatomic molecular band spectra may be observed. The band spectra produced by various materials in a discharge have long been studied and the wavelengths catalogued. The work of PEARSE and GAYDON (1963) is well known.

Organic compounds containing C and H atoms produce primarily C_2 band, CH band, H line and C line spectra. CN bands are observed, primarily as a result of nitrogen impurity in the carrier gas but also arise from nitrogen-containing organic compounds.

Sulfur dioxide and carbonyl sulfide produce S_2, SO bands and S and O line spectra in the helium discharge. The S_2 and SO bands are extensive and somewhat diffuse. Ammonia produces chiefly H line and NH bands. Nitrogen oxides produce O lines and N lines. The wavelengths of the main bands or lines of interest in detection of air pollutant compounds are given in TABLE 1.

Air pollutant compounds selected for study included carbon monoxide, carbon dioxide, the nitrogen oxides, sulfur dioxide, and carbonyl sulfide. Alcohols and some others were selected as representative organic compounds.

Limits of detection for the inorganic gases were sought as they had not been previously reported. The technique of multiple readout was explored to determine if selectivity of detection could thereby be improved.

* Presented at the Symposium on Advances in Instrumentation for Air Pollution Control, sponsored by National Air Pollution Control Administration, U.S. Department of Health, Education and Welfare, Cincinnati, Ohio, 26–28 May 1969.

TABLE 1. SPECTRAL EMISSION WAVELENGTHS OF AIR POLLUTANT COMPOUNDS

Compound	Wavelength (Å)	Band or line	Remarks
SO_2	3064.1	SO	Degraded to the red
	3164.8	SO	
	3271.0	SO	
	7771–7774	O	Oxygen triplet
	2769.4–6165.8	S_2	These bands merge with the SO bands to produce almost a continuum of background in discharges.
CO	5165.2	C_2	Most intense of C_2 series
	7771–7774	O	Oxygen triplet only if O_2 impurities are low
CO_2	5165.2	C_2	
	7771–7774	O	Oxygen triplet, stronger than for CO
COS	5165.2	C_2	
	3064.1	SO	Also other SO bands
	7771–7774	O	Less prominent
N_2O, NO, NO_2	3883.4	CN	
	4216.0	CN	Best, avoids interference from He line at 3888 Å
	6008.5	N	Other N lines are likely useful
	4358.3	N	
O_2	7771–7774	O	Good selectivity
H_2	6562	H	Good selectivity
Organic compounds in general	5165.2	C_2	
	4314.2	CH	
	3883.4	CN	
	6562	H	

EXPERIMENTAL APPARATUS

A single channel d.c. discharge detector of the design reported by BRAMAN (1968) was constructed and used with a Heath Company scanning monochrometer. A Tracor Instrument Co., Microtek MT-220 gas chromatograph was used where separations and analysis were desired. Helium carrier gas was partially purified by means of a 24 ft × $\frac{3}{8}$ in. dia. column of activated charcoal cooled to liquid nitrogen temperature.

A four-channel detector was also constructed employing four interference filters at selected line and band wavelengths instead of a monochrometer. FIGURE 1 shows the apparatus arrangement when the four channel detector was used. Photomultiplier (PM) tubes used were 1P28 for the C_2(5165Å), CH(4512Å) and CN(4216Å) bands observed. A S1 response tube, 7102, was used with the H-line filter (6562Å).

The d.c. discharge power supply was a simple voltage doubler circuit with a Π filter. Cathode follower circuits were used with the PM tubes. The PM tube power supply was constructed using voltage regulator tubes. Strip chart recorders of commercial manufacture were used.

Carrier gas flow rates were 90–120 ml min^{-1}, optimum for the d.c. discharge

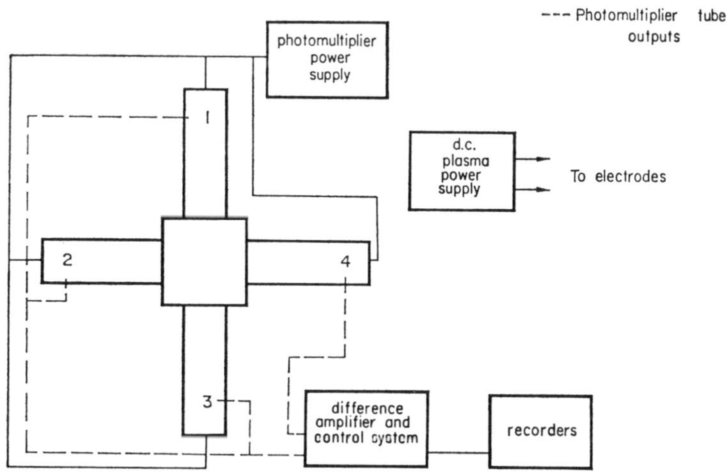

FIG. 1. Component parts of a four-channel DCD detector.

response. The d.c. discharge was operated with a $\frac{1}{2}$ in. gap and at an optimum power level of 35 W in^{-1}.

Chromatographic columns used included a 13 × molecular sieve column, 3 ft × $\frac{1}{8}$ in. o.d., a 10 per cent Carbowax 20M on Chromosorb W, $2\frac{1}{2}$ ft × $\frac{1}{4}$ in. o.d., and a 5 per cent SE30 on Chromosorb W, 3 ft × $\frac{1}{8}$ in. o.d. Temperatures ranged from room temperature to 100°C.

Sample handling

Gas samples were introduced into the gas chromatograph by means of a gas sampling valve with approximately 10 μl sample volume loop. An exponential dilution flask was used to generate a variety of gas sample concentrations. A glass stop cock sample bypass system was also used in the analysis of CO and H_2 mixtures at reduced pressures.

Liquids were sampled by preparing solutions in much higher boiling or much lower boiling solvents. Amounts of material passing through the d.c. discharge were limited to 0.1–0.5 μl per sample.

Data handling

Data reduction from the four channel detector was initially done by hand integration of peak areas. Response curves were then calculated in the usual manner. Data obtained when the exponential dilution flask was used were processed by computer using a Fortran program.

The program calculated the sampled gas concentrations from the exponential dilution flask, the response line (regression analysis) and standard deviation of the response line. Data for the Fortran program was obtained by hand integration of recorder charts, punched on cards and then run through the computer program.

Selected gases

The emission spectra of several selected gases in the d.c. discharge were studied using the Heath monochrometer. The most intense emission bands found corresponded, as expected, with those in TABLE 1. Sulfur dioxide and carbonyl sulfide produce S_2, SO bands and S and O line spectra in a helium discharge. The S_2 and SO bands are extensive and somewhat diffuse. Nitrogen oxides produce O lines and N lines.

Limits of detection for the nitrogen oxides, and other permanent gases were determined from responses at C_2, CH, CN, H lines and N or O line wavelengths. Data are given in TABLE 2. The notation "quenches" indicates that background radiation at the indicated wavelength is reduced by the sample component.

TABLE 2. LIMITS OF DETECTION FOR SELECTED GASES BAND OR LINE SPECTRA OBSERVED

Gas	C_2	CH	CN	H	Others
N_2	1% of CN	10% of CN	1.4×10^{-10} l.	3% of CN	N line
O_2	quenches	quenches	none	none	O line
H_2	none	none	quenches	9.5×10^{-11} l.	—
NO	none	none	3.3×10^{-9} l.	quenches	O line N line
N_2O	none	20% of CN	1×10^{-9} l.	quenches	N line O line
NO_2	30% of CN	none	4×10^{-10} l.	none	O line N line
CO	4.2×10^{-11} l.	25% of C_2	very weak	very weak	O line
CO_2	2.9×10^{-10} l.	2.9×10^{-10} l.	quenches	quenches	O line
COS	7×10^{-10} l.	7×10^{-10} l.	quenches	weak	O line
SO_2	none	very weak	weak	none	O line
CF_4	30% of CH	7×10^{-10} l.	quenches	very weak	—

Based upon the analysis of 1 ml samples of gas, the limits of detection are in the 0.1–1 ppm (vol.) range for the permanent gases. For nitrogen oxides this method is not so sensitive as the chemiluminescent techniques (FONTION, SABADELL and RONCO, 1970) where limits of detection are approximately 4 ppb. The d.c. discharge technique does have a lower limit of detection for SO_2 than the WEST and GAEKE method (1956) which is suitable for analysis of microgram amounts. Major qualitative differences are found between the gases by comparing response at the several wavelengths.

Carbon containing gases all give good positive responses at C_2 and CH band wavelengths while the nitrogen oxides do not.

Limits of detection indicated the d.c. discharge detector was suitable for gas analysis of automobile exhaust. Several exhaust samples were examined. The four channels of the detector were found particularly useful in the development part of the analytical work to confirm the identity of air peaks, hydrogen, CO, hydrocarbons and CO_2.

Organic compounds

The multiple response qualitative identification technique. The four emission wavelength channels are useful for qualitative identification of eluted components in a gas chromatographic system if response curves are known. It has been established by BRAMAN (1968) that emission response ratios depend upon the elemental make up and structure of compounds producing the response. Simultaneous observation of four channels produces six ratios of responses and thus six values for qualitative identification. If responses are obtained in the linear region, at some particular wavelength, then, simply:

$$S_{\lambda} = \underline{a} K_{\lambda_1} g$$

where $S_{\lambda 1}$ is the response observed at wavelength λ_1 (the peak area or deflection at some time during elution of the component), g is sample weight, K_{λ_1} is an instrument constant; $\underline{a}$ is the constant expressing the degree to which a compound produces the excited atom or radical detected at λ_1. The constant, K_{λ_1}, contains signal gain factors from the instrument employed and remains constant unless factors such as PM tube voltage, slit width or the interference filter is changed. The value of $\underline{a}$ will be influenced by the nature of the electrical discharge but is apparently constant up to 3–5 μg of sample size. The value of K_{λ_1} will be different from one instrument to another because the electrical and optical systems will not be identical. Thus the raw data response ratios, S_{λ_1} S_{λ_2} are qualitatively of use only in an individual instrument and not between instruments.

A process of normalization of response ratios to one or several standard compounds eliminates the problem of differences between instruments. Selecting ethanol as a standard the following develops in comparison to propanol.

$$S_{\lambda_1}\text{ (ethanol)} = a_{\lambda_1} K_{\lambda_1} g$$

$$S_{\lambda_2}\text{ (ethanol)} = a_{\lambda_2} K_{\lambda_2} g$$

$$S_{\lambda_1}\text{ (propanol)} = a^1_{\lambda_1} K_{\lambda_1} g^1$$

$$S_{\lambda_2}\text{ (propanol)} = a^1_{\lambda_2} K_{\lambda_1} g^1$$

$$\frac{S_{\lambda_1}\text{ (ethanol)}}{S_{\lambda_2}\text{ (ethanol)}} = \frac{a_{\lambda_1}}{a_{\lambda_2}} \frac{K_{\lambda_1}}{K_{\lambda_2}} \text{ (response ratio for ethanol)}$$

$$\frac{S_{\lambda_1}\text{ (propanol)}}{S_{\lambda_2}\text{ (propanol)}} = \frac{a^1_{\lambda_1}}{a^1_{\lambda_2}} \frac{K_{\lambda_1}}{K_{\lambda_2}} \text{ (response ratio for propanol).}$$

By normalizing to ethanol

$$\frac{(\text{propanol})\ \dfrac{a^1_{\lambda_1}.K_{\lambda_1}}{a^1_{\lambda_2}.K_{\lambda_2}}}{(\text{ethanol})\ \dfrac{a_{\lambda_1}\ K_{\lambda_1}}{a_{\lambda_2}\ K_{\lambda_2}}} = \frac{a^1_{\lambda_1}/a^1_{\lambda_2}\ (\text{propanol})}{a_{\lambda_1}/a_{\lambda_2}\ (\text{ethanol})} = N.$$

The normalized response ratio, N, is independent of sample size and the instrument factors. With a published set of N values available it should be necessary to run only a normalizing standard compound for calibration of the instrument response. Unknown peaks are then identified by calculating normalized response ratio values and by comparing to a known list. Six N values are available for each compound from a four-channel detector.

In order to investigate the use of the normalized response ratios an alcohol series, methanol through *n*-nonanol, was studied. Samples of the alcohols were passed through a gas chromatography column (Carbowax) and the responses determined.

TABLE 3. EMISSION RATIOS FOR AN ALCOHOL SERIES

	CH/C_2	CH/CN	CH/H	C_2/CH	C_2/H	CN/H
Methanol	1.37	6.17	3.47	4.47	2.53	0.56
Ethanol	1.21	5.86	3.70	4.78	3.07	0.64
n-Propanol	1.00	5.49	3.66	5.49	3.66	0.67
n-Butanol	0.80	5.23	5.42	6.47	6.78	0.68
n-Pentanol	0.78	5.03	5.72	6.42	7.28	1.13
n-Hexanol	0.76	5.17	5.89	6.86	7.79	1.14
n-Heptanol	0.73	4.93	6.00	6.76	8.23	1.21
n-Octanol	0.68	4.82	6.54	7.10	9.58	1.35
n-Nonanol	0.64	4.62	6.00	7.17	9.35	1.30
2-Propanol	0.75	5.06	4.47	6.69	5.96	0.89
iso-Butanol	0.68	4.68	6.24	6.87	9.17	1.34

94 ml He min^{-1}.
35 W in^{-1}.
0.50 in. plasma gap.

TABLE 3 gives emission ratios obtained. Data were obtained from responses of three or more different sample sizes of the indicated compound. All sample sizes were in the linear response range. Normalized emission response ratios were calculated using ethanol as a base and are plotted versus carbon number in FIG. 2. A regular change in the normalized ratios is observed for most of the values. Deviations may be due to a few poor response values. For example, the H-line response for the *n*-C_8 alcohol is likely in error on the low side. Iso-compounds gave a considerably different set of N values than the corresponding normal alkanol. Repeatability of the normalized response values is approximately $\pm$ 5 per cent.

This series substantiates the multiple response qualitative identification technique. Taken together, the six ratios are a unique set for each compound. TABLE 4 gives

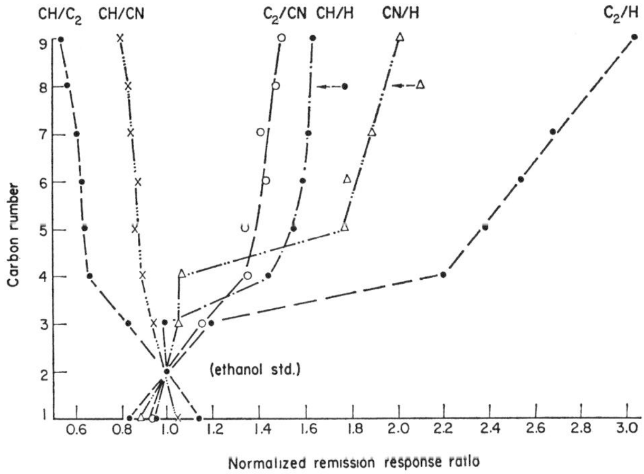

FIG. 2. Normalized emission response ratios as a function of carbon number (*n*-alkanol series).

normalized data for other compounds. Differences between benzene, toluene and m-xylene are small for the three ratios given. The values are significantly different from those for the alcohols.

TABLE 4. NORMALIZED EMISSION RATIOS FOR SELECTED ORGANIC COMPOUNDS

	CH/C_2	CH/CN	C_2/CN
Cyclohexane	1.13	0.53	0.47
Benzene	0.74	0.29	0.40
Toluene	0.74	0.31	0.43
m-Xylene	0.75	0.30	0.40
Ethanol	1.00	1.00	1.00

* Data for the H channel were not obtained due to inoperability of that circuit during these experiments.

Further work in developing *N* values for a variety of compounds encountered in pollution studies is anticipated.

Acknowledgement—The author wishes to acknowledge the support given by the National Science Foundation, Grant No. GP 8598, under which the work was done.

REFERENCES

BRAMAN R. S. and DYNALCO A. (1968) *Anal. Chem.* **40,** 95.

FONTION A., SABADELL A. J. and RONCO R. J. (1970) *Anal. Chem.* **42,** 575.

MCCORMACK A. J., TONG S. C. and COOKE W. D. (1965) *Anal. Chem.* **37,** 1470.

PEARSE R. W. B. and GAYDON A. G. (1963) *The Identification of Molecular Spectra*, Ugsman & Hall Ltd.

WEST P. W. and GAEKE G. C. (1956) *Anal. Chem.* **28,** 1816.

Detection of Air Pollutants with Tunable Diode Lasers

E. D. Hinkley and P. L. Kelley

In this article we describe the use of tunable $Pb_{1-x}Sn_xTe$ semiconductor diode lasers in remote or long-range sensing and point sampling of molecular pollutant gases. Remote sensing can be accomplished by means of measurements of the infrared absorption or emission lines characteristic of a particular pollutant at atmospheric pressure. Point sampling permits gas pressures to be reduced until the Doppler-broadened infrared absorption structure is revealed, thereby making possible very high specificity. Since one measurement technique is used for all the pollutant gases, the operational requirements are simplified. Moreover, the essentially instantaneous response is advantageous for a number of applications.

Several optical techniques for the detection of pollutant gases are now being explored in various laboratories. They range from classical methods in which selective filters, spectrographs, interferometers, and optical-correlation instruments are used (*1*) to the use of lasers for direct absorption (*2*) and Raman scattering measurements (*3*). One advantage of optical sensing (*4*) over conventional wet chemical methods is that one can perform the measurements in situ by measuring emission spectra, by detecting scattered laser radiation, or by making transmission measurements over relatively long atmospheric paths. Infrared spectroscopy is considered an important technique since each of the major molecular pollutants contains at least one strong band in the region from 3 to 15 micrometers. Dispersive infrared instruments now being used in the field usually have resolutions that are much poorer than the linewidths of typical gases at atmospheric pressure, and thus potentially useful information is lost. Even when high-quality spectrometers are used, the resolving power must usually be reduced in order to maintain a reasonable signal-to-noise ratio with the relatively low power per unit spectral range available from incoherent sources. The use of gas lasers as nondispersive instruments, on the other hand, is severely restricted by the fact that these lasers cannot be tuned appreciably from their nominal wavelengths, and the match to pollutant gas absorption lines is seldom ideal (*2*).

Semiconductor lasers of $Pb_{1-x}Sn_xTe$ can be "tailored" chemically, by an adjustment of the composition factor x, to emit anywhere in the wavelength range from 6.5 to 32 micrometers (*5*), thereby permitting an ideal match to the strong infrared absorption lines of most of the molecular pollutant gases. Individual lead-salt lasers can be tuned by applying pressure (*6*), by applying an external magnetic field (*7*), or by varying the laser temperature (*8*). In a simple manifestation of temperature-tuning, a $Pb_{1-x}Sn_xTe$ diode laser can be tuned over a range of 40 reciprocal centimeters quasi-continuously (from one continuously tunable longitudinal mode to another) by changing the diode curent. This tuning range is sufficient to permit identification of the spectral signature of a given pollutant even in the presence of other molecular species, and to permit a quantitative determination of its concentration. These features overcome the major limitations of conventional gas lasers in monitoring applications.

Other tunable laser sources may also be useful for air pollution monitoring. These include the spin-flip Raman lasers (*9*), organic dye lasers, and systems involving parametric oscillators or difference frequency generators. At the present time these devices are much more complex than the tunable semiconductor diode lasers; nevertheless, their wide tuning ranges should soon find many applications in spectroscopy.

Tunable Diode Lasers

The diode lasers are fabricated from vapor-grown single crystals of Pb_{1-x}-Sn_xTe semiconducting material (*10*) by cleaving the crystals into rectangular parallelepipeds (each containing a *p-n* junction) of approximate overall dimensions 0.12 by 0.05 by 0.03 centimeter. Low-resistance ohmic contacts are formed, and the laser is then mounted onto the cold-finger of a cryogenic Dewar. Liquid helium temperatures are presently required for continuous laser operation, with pulsed operation possible at liquid nitrogen temperature (77°K). The infrared emission frequency of the laser is, within the spontaneous emission bandwidth, determined by the refractive index of the semiconductor and the physical length of the cavity. This frequency can be fine-tuned by a small change in the diode current (*11*), which alters the refractive index of this mixed-crystal semiconductor (which has low thermal conductivity) through heating. A single diode laser can be "tuned" in this manner over 40 reciprocal centimeters in continuous bands up to 2 reciprocal centimeters wide; but because of a change in laser operation from one cavity mode to another, emission occurs at only about one-half of the frequencies in this tuning range. Although continuous tuning is essential for fundamental high-resolution spectroscopic studies (*12*), the broader, quasi-continuous tuning is quite adequate for application to the detection of pollutant gases. That the diode laser provides effectively infinite resolution for these applications was demonstrated by a heterodyne experiment involving a carbon dioxide laser (*13*), in which we showed that the linewidth of a 0.24-milliwatt $Pb_{0.88}Sn_{0.12}Te$ diode laser at 10.6 micrometers was 54 kilohertz (1.8×10^{-6} reciprocal centimeter). By comparison, the narrowest Doppler-broadened linewidths of even the heaviest molecular pollutants are tens of megahertz at room temperature.

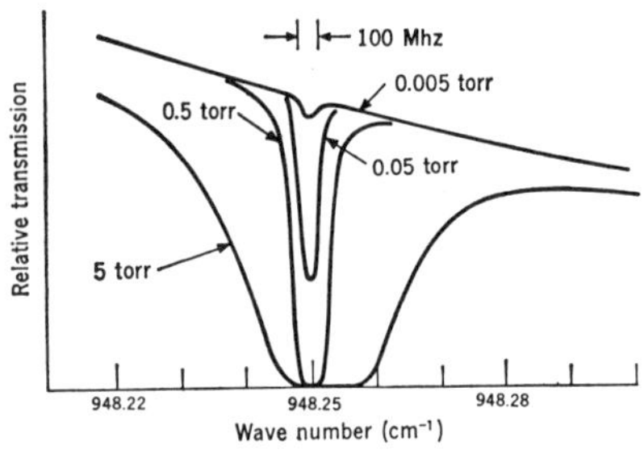

Fig. 1. Absorption spectra of the $sP(1,0)$ line of ammonia taken at various gas pressures with a tunable $Pb_{0.88}Sn_{0.12}Te$ diode laser. Cell length, 10 centimeters; temperature, 300°K. Resolution (diode laser linewidth) is better than 0.1 megahertz (3.3×10^{-6} reciprocal centimeter).

Application to Pollutant Detection

In a typical monitoring application, several $Pb_{1-x}Sn_xTe$ diode lasers would be used, each tailored to emit in a strongly absorbing infrared region of one of the pollutant gases. This arrangement is necessary because of the limited current-tuning range of these lasers, and desirable because it provides for increased specificity. The small size of these lasers makes it possible for several to be mounted into the same cryogenic Dewar. Specific applications to pollutant detection can be divided into three categories: point sampling, long-path atmospheric transmission, and remote emission sensing.

Point sampling. Use of this technique permits the absorption spectra to be measured at reduced pressure where possible overlap with neighboring spectral lines of the gas (or with other gases in the atmospheric sample) is minimized. Several experiments have been performed with $Pb_{0.88}Sn_{0.12}Te$ diode lasers in the 10.6-micrometer region to illustrate their usefulness in such an application. Figure 1 shows absorption profiles of the strong $sP(1,0)$ line of ammonia at several pressures. At low pressure the full linewidth at half peak absorption is seen to approximate the Doppler-limited value of 85 megahertz. As the ammonia pressure is increased to 0.5 torr, the cell (10 centimeters long) becomes essentially opaque to radiation near the line center. The improvement in sensitivity of this nondispersive technique can be appreciated if a comparison is made between this spectrum and one obtained with a grating spectrometer having a resolution of 0.1 to 0.2 reciprocal centimeter (*14*). With such a grating spectrometer an ammonia pressure of 5 torr in a cell 200 centimeters long produced the same maximum absorption as that shown here for the curve at a pressure of 50×10^{-3} torr where the amount of ammonia in the optical path is 2000 times smaller.

Figure 2 shows the ν_3 vibration-rotation band of sulfur hexafluoride obtained with a high-resolution grating spectrometer (*15*), as well as a segment of this band taken with a tunable $Pb_{0.88}Sn_{0.12}Te$ diode laser. Since sulfur hexafluoride is not usually present in the atmosphere, except near leakage sources such as high-voltage transformers, it can be used as a trace gas in certain monitoring applications (*16*). Sulfur hexafluoride has an exceedingly complex spectrum in the infrared region, which, as shown in Fig. 2, cannot be resolved in the 950–reciprocal centimeter region even with a high-quality grating spectrometer, but which is resolved by the diode laser scan. The widths of these lines correspond to the 29-megahertz value predicted for the Doppler-broadening of sulfur hexafluoride at room temperature, as compared with 85 megahertz for ammonia. Since a determination of mass can be made from the Doppler width, we can use tunable lasers to distinguish between gaseous species on the basis of their molecular weights, and thereby confirm pollutant identification by point sampling at reduced pressure. The abscissa of Fig. 2 was precisely calibrated relative to the *P*(16) carbon dioxide laser line by viewing on a spectrum analyzer the wide-band heterodyned signal between this line and that from a tunable diode laser, with an absorption cell in the path of the diode laser beam (*12*).

Ethylene is one of the many hydrocarbons present in automobile exhaust. It is not usually monitored separately, but is lumped with the other hydrocarbons in a single measurement of total hydrocarbons. Figure 3A shows the absorption spectrum of ethylene near the *Q* branch of the ν_7 vibration-rotation band. The 25 percent drop in laser intensity near 949.24 reciprocal centimeters is produced by a shift of laser power from this longitudinal mode (*17*). By modulating the diode current with a small (0.1 percent) superimposed sinusoidal current at a frequency of a few hundred hertz, the laser emission frequency is modulated, thus allowing synchronous detection to be obtained without mechanical chopping, and, in addition, yielding the derivative of the absorption spectrum. This is shown in Fig. 3B where the top trace is a calibration spectrum of ethylene [1000 parts per million (ppm) in nitrogen]. A total gas pressure of 5 torr produces a maximum derivative signal for ethylene, representing a balance between pressure-broadening (upper limit) and a reduced total number of absorbing molecules (lower limit). The three lower traces represent spectra obtained from automobile exhaust specimens, also taken at a pressure of 5 torr, but with increased amplifier gain. By measuring the signal amplitudes and comparing them with the calibration curve, ethylene concentrations of from 22 to 74 ppm are deduced. It is unmistakably clear that each automobile exhaust sample contains ethylene, and that there is no interference from water vapor or any other molecular constituents present. Moreover, the point at which the diode laser power changed abruptly is easily identifiable, and this power change affects neither the qualitative nor the quantitative comparisons.

There is no question that this technique is very specific at low pressures where Doppler-broadening is predominant and collision-broadening is negligible. Here each line is about 6×10^7 hertz (2×10^{-8} reciprocal centimeter) wide. The total (compositional) tuning range of the $Pb_{1-x}Sn_xTe$ diode lasers is about 3×10^{13} hertz. If we assume 50 gaseous molecular species in the atmosphere, each with 2000 reasonably strong lines in this range, the separation between individual lines should average about five times the linewidth.

The maximum absorption constant per unit pressure of the strongest lines of ammonia, sulfur hexafluoride, and ethylene approximate 2 reciprocal cen-

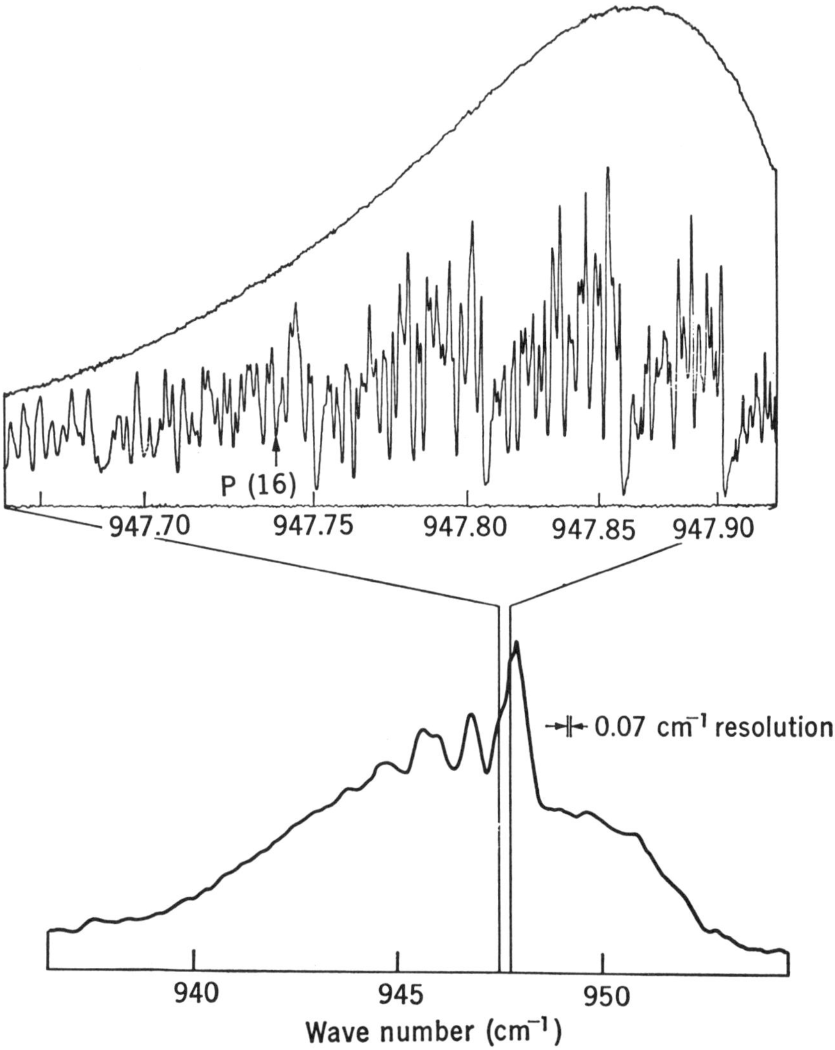

Fig. 2. (Bottom) Grating spectrometer absorption curve of the ν_3 band of sulfur hexafluoride at room temperature [adapted from the curve of Brunet and Perez (*15*)]; sulfur hexafluoride pressure, 0.1 torr; cell length, 25 centimeters. (Top) Diode laser transmission scan of a segment of this band near the *P*(16) carbon dioxide laser line; sulfur hexafluoride pressure, 0.1 torr; cell length, 10 centimeters; resolution, 3×10^{-6} reciprocal centimeter.

timeters per torr. A detection limit of 0.3 ppm is expected for such lines in a cell 1 meter long on the basis of these initial experiments; but this limit should improve to the low parts-per-billion (ppb) range with the use of longer path-length multireflection cells and improved electronics. Moreover, noting that the strengths of strong absorption lines of these three quite different gases are nearly equal, we expect those for a number of pollutant gases to have the same general values, thus yielding ultimate detection limits that are also in the low parts-per-billion range.

Long-path atmospheric transmission. Even at atmospheric pressure the collision-broadened widths of spectral lines for typical gases are well below the resolution limit of infrared spectrometers now available for field use. If a tunable diode laser is used, the resolution limit is imposed by the pressure-broadened linewidth of the gas itself, to yield all the available optical information within the tuning range. Monitoring of pollutant gases in the atmosphere by long-path transmission has the advantage over point sampling in that it produces an *average* value for the pollutant gas concentration over the path, which may be of greater overall significance in ambient air analysis.

If we assume that the pressure-broadened linewidth of the pollutant gas equals the Doppler width at about 5 torr, and if we use a Doppler-broadened absorption constant per unit pressure of 2 reciprocal centimeters per torr, the absorption constant α (in reciprocal centimeters) at the center of a strong line at atmospheric pressure is

$$\alpha \approx 10^{-5} c \qquad (1)$$

where c is the concentration in parts per million. Over a 1-kilometer path, a pollutant gas with an average concentration of 1 ppb will produce a 0.1 percent change in transmission as we tune through this absorption line.

We shall now consider the collection efficiency. The emitting area of the diode laser (which operates in a low-order transverse mode) is 500 by 40 micrometers. If a $\times 20$ beam expander is used, a laser beam near a wavelength of 10 micrometers diverges to an area of about 10^5 square centimeters 1 kilometer from the source. For a collection area of 100 square centimeters and a 1-watt (peak) pulsed diode laser (tuned by varying either the pulse width or the repetition rate), the collected power in the absence of attenuation is 1 milliwatt. For an average pollutant concentration of 1 ppb over this path, a signal change of 1 microwatt is produced, and this change can be measured easily by the use of synchronous detection techniques. A corner reflector (of the order of 100 square centimeters or greater area to avoid diffraction losses in the return beam) situated remotely would eliminate the requirement of having instrumentation at two locations. By positioning a number of corner reflectors at various points around the source, many radial measurements could be made from the same monitoring station. It should also be pointed out that heterodyne detection of absorption lines, with an offset local oscillator, should yield an improved signal-to-noise ratio because of the narrow emission linewidth of the laser.

Transmission experiments should be performed with these diode lasers, operating both pulsed and continuous, to test the feasibility of long-path monitoring, and to show whether or not specificity in identifying a particular gaseous pollutant can be retained in the presence of absorption lines from other molecular constituents. As is well known, the high abundances of water vapor and carbon dioxide in the atmosphere can cause serious interference with infrared spectroscopy. We therefore restrict ourselves to considering the atmospheric "window" between 8 and 13 micrometers, which is approximately 500 reciprocal centimeters wide. For example, lasers tailored to emit in the 8.6-micrometer region are better suited to sulfur dioxide detection than those tailored to the (somewhat stronger) 7.5-micrometer absorption band. Even in the "window" region, absorption in the wings (many linewidths away from the line center) of water vapor and carbon dioxide lines may be greater than that corresponding to pollutants in small concentrations; however, this effect is somewhat compensated by the fact that the slopes of these wings are relatively flat in relation to the slope of a pollutant absorption line near its peak. Derivative spectroscopy, similar to that shown in Fig. 3B, will improve detection of these small peaks relative to the slowly varying Lorentzian wings of the dominant atmospheric absorption lines.

We shall now consider the interference between pollutant molecules at atmospheric pressure and start with the assumption that there are 25 molecular species present that absorb in the "window" between 8 and 13 micrometers. At atmospheric pressure the lines are broadened to about 6×10^9 hertz, and most of the structure within individual J values of the P and R branches disappears; the Q branch, if present, also loses most of its structure since closely spaced strong lines overlap and weak lines become lost in the tails of strong lines. We estimate, therefore, that at atmospheric pressure there are only 100 resolvable lines per molecule; this estimation implies that, on the average, the peaks of the lines are separated by one linewidth. On the assumption that each line is resolvable if a frequency interval three linewidths wide ($\sim 18 \times 10^9$ hertz) is not occupied by an absorption line from another molecule, and that there is a uniform probability distribution, there should be an average of five resolvable lines per molecular species.

Remote emission detection. One of the more difficult problems in air pollution monitoring is gaining access to industrial smokestacks. Optical remote-sensing techniques can be extremely useful for such applications.

If the $Pb_{1-x}Sn_xTe$ diode laser is used as a tunable local oscillator, infrared heterodyne detection techniques (*18*) may be used to detect spectral emission lines from pollutants present in smokestack effluent (*19*). (The technique is similar to that used by radio astronomers to detect weak emissions from distant sources.) For heterodyne detection with an ideal photon counter (unit quantum efficiency and zero excess noise), the signal divided by the local oscillator-generated noise is given by

$$\frac{S}{N} = \frac{1}{h\nu}\left(\frac{\partial P_s}{\partial \nu}\right) (B\tau)^{1/2} \qquad (2)$$

where $\partial P_s/\partial \nu$ is the received signal power per unit frequency range in a single spatial mode, $h\nu$ is the infrared photon energy, B is the intermediate-frequency (IF) bandwidth of the detection system, and τ is the postdetection integration time. For the case where the solid angle subtended by the source at the detector is larger than the diffraction-limited solid angular resolution of the collection optics—a situation usually valid for smokestack

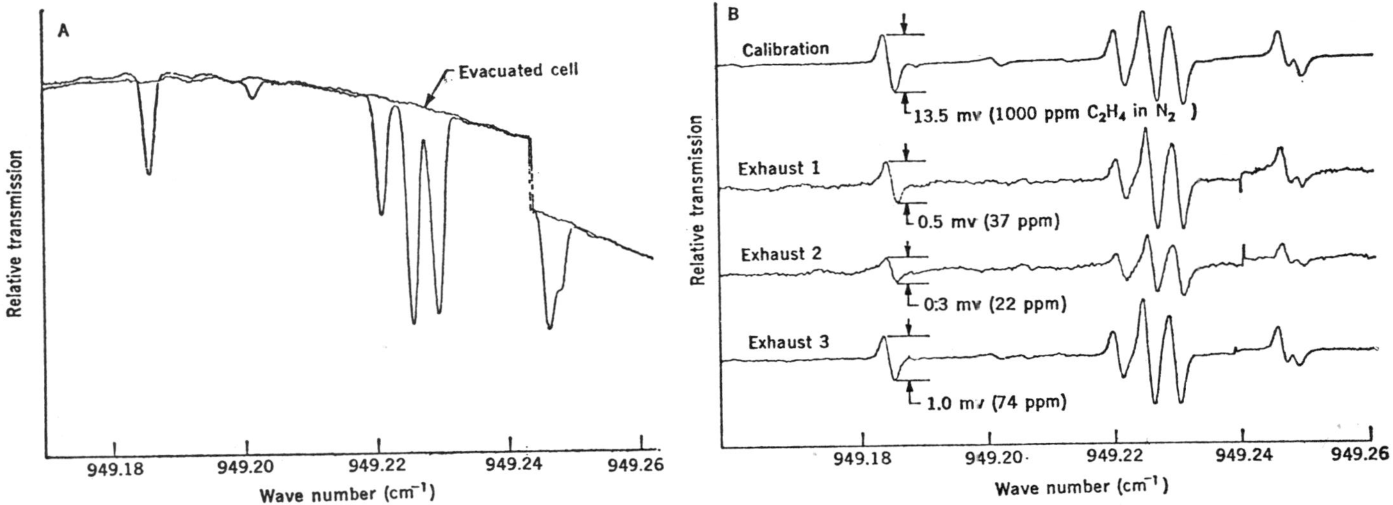

Fig. 3. (A) Absorption spectrum near the *Q* branch of ethylene at a pressure of 25×10^{-3} torr; cell length, 30 centimeters. See (*17*) for experimental procedure. (B) Derivative spectra of a calibration sample of ethylene (1000 ppm in nitrogen), and three automobile exhaust samples at different amplifier gain settings. Total pressure, 5 torr; cell length, 30 centimeters.

plumes at ranges up to 1 kilometer and with optics as small as 1 square centimeter—the signal-to-noise ratio becomes

$$\frac{S}{N} = (1 - e^{-\alpha L})\,(B\tau)^{1/2} \times \left[\frac{1}{\exp\frac{h\nu}{kT_s} - 1} - \frac{\epsilon_b}{\exp\frac{h\nu}{kT_b} - 1}\right] \quad (3)$$

where L is the thickness of the plume and T_s is its temperature, ε_b is the emissivity of the background at temperature T_b, and k is Boltzmann's constant. We have assumed: (i) that the IF bandwidth is less than the emission linewidth; (ii) that the emission and absorption due to the pollutant in the ambient atmosphere is not important because of the low concentration of the pollutant in the atmosphere relative to that in the plume; (iii) that the background attenuation from the wings of other molecular absorption lines is negligible; and (iv) that the local oscillator has sufficient power to overcome the other sources of noise (this requirement is readily satisfied with a 1-milliwatt diode laser). Since ε_b is often much smaller than unity (about 0.1 in the 10-micrometer region of the infrared when there are no clouds in the background), the source does not have to be hotter than the background; in fact, it could be colder. If we choose the following representative values for the parameters: $B = 10^8$ hertz, $\tau = 1$ second, $\alpha = 10^{-5}\ c$ reciprocal centimeters (where c is the concentration in parts per million), $L = 1$ meter, $h\nu =$ 0.125 electron volt, T_s = 400°K, T_b = 300°K, and $\varepsilon_b = 1$, we find that

$$\frac{S}{N} \approx 0.2\ c$$

It should be possible, therefore, to carry out remote detection of pollutant molecules such as sulfur dioxide in smokestack plumes in concentrations greater than 10 ppm. In order to make absolute measurements of concentration, αL must be less than unity and the gas temperature must be known. If αL exceeds unity, so that the plume is optically dense, then a weaker emission line can be selected. Gas temperature can be measured in several ways: (i) by heterodyne detection of one of the stack gases at two wavelengths; (ii) by multicolor broadband radiometric measurements; or (iii) by direct measurement at the top of the stack.

Atmospheric turbulence may cause fluctuations in the received signal. This fluctuation can be overcome by alternately tuning the local oscillator between the peak emission wavelength of the pollutant gas and a wavelength of negligible emission in a time short enough so that turbulent effects are negligible (~ 1 millisecond), and then summing the excess pollutant signal for a large number of measurement intervals. Two local oscillators could also be used simultaneously, one operating at the emission peak and the other offset from it.

General Considerations

The absorption bands of a few of the major molecular pollutant gases (such as carbon monoxide and nitric oxide) are beyond the compositional tuning range of $Pb_{1-x}Sn_xTe$ diode lasers (*20*). Development of other semiconducting materials, for example, $PbS_{1-x}Se_x$, which has lased with an electron beam as pump source (*21*), may yield tunable diode lasers useful for detecting these gases as well. Semiconductor lasers that operate in the region from 3 to 6.5 micrometers should have the added advantage of a higher ultimate operating temperature.

Although temperatures below 20°K are presently required for continuous operation, pulsed laser emission can be obtained at liquid nitrogen temperature (77°K). With the availability of relatively inexpensive cryogenic coolers, it appears possible to achieve fast on-line monitoring of gases present in automobile exhausts or in chemical processing plants by direct absorption techniques with the use of diode lasers, each of which is tailored to a strong absorption line of one of the gases to be monitored. It is expected that further improvements in materials fabrication will eventually permit continuous operation at this temperature as well.

Summary

Preliminary experiments indicate that tunable $Pb_{1-x}Sn_xTe$ diode lasers will be useful in the identification and sensitive detection of most of the atmospheric pollutant gases. For point-sampling applications, concentrations in the parts-per-billion range should be measurable with very high specificity. For long-range atmospheric transmission techniques, the improved resolution capability and tunability of these diode lasers make them attractive replacements for spectrometers and fixed-frequency laser sources where operation at cryogenic temperatures is not a serious impediment. By using these lasers as tunable local oscillators in the infrared heterodyne configuration, remote passive detection of gases present in smokestack effluent appears possible. Finally, pulsed operation at temperatures available with simple cryogenic coolers permits immediate application to the fast detection of gases present in automobile exhaust and in chemical processing plants.

References and Notes

1. A. R. Barringer, *J. Opt. Soc. Amer.* **60**, 729 (1970); C. B. Ludwig, M. Griggs, E. R. Bartle, *ibid.*, p. 731.
2. H. J. Gerritsen, *Trans. Amer. Inst. Mining Eng.* **235**, 428 (1966); G. B. Jacobs and L. R. Snowman, *Inst. Elec. Electron. Eng. J. Quantum Electron.* **QE-3**, 603 (1967); D. N. Jaynes and B. H. Beam, *Appl. Opt.* **8**, 1741 (1969); P. L. Hanst, *Appl. Spectrosc.* **24**, 161 (1970); R. T. Menzies, N. George, M. L. Bhaumik, *Inst. Elec. Electron. Eng. J. Quantum Electron.* **QE-6**, 800 (1970).
3. H. Inaba and T. Kobayasi, *Nature* **224**, 170 (1969); T. Kobayasi and H. Inaba, *Appl. Phys. Lett.* **17**, 139 (1970).
4. Microwave spectroscopy has many of the advantages of optical spectroscopy for the detection of molecular pollutants, but with the following limitations: (i) insufficient resolution at atmospheric pressure; (ii) very low pressures required for high-resolution studies in the Doppler-limited regime; and (iii) molecules such as carbon doxide, ethylene, and benzene, which lack a dipole moment in the ground state, cannot be detected.
5. J. O. Dimmock, I. Melngailis, A. J. Strauss, *Phys. Rev. Lett.* **16**, 1193 (1966); J. F. Butler, A. R. Calawa, T. C. Harman, *Appl. Phys. Lett.* **9**, 427 (1966); J. F. Butler and T. C. Harman, *ibid.* **12**, 347 (1968); for a review, see I. Melngailis, *J. Phys. Colloq.* **C-4** (suppl.), No. 11–12 (1968), pp. C4–84 to C4–94; T. C. Harman, *J. Phys. Chem. Solids*, in press.
6. J. M. Besson, J. F. Butler, A. R. Calawa, W. Paul, R. H. Rediker, *Appl. Phys. Lett.* **7**, 206 (1965); J. M. Besson, W. Paul, A. R. Calawa, *Phys. Rev.* **173**, 699 (1968).
7. J. F. Butler and A. R. Calawa, in *Physics of Quantum Electronics*, P. L. Kelley, B. Lax, P. E. Tannenwald, Eds. (McGraw-Hill, New York, 1966), p. 458; A. R. Calawa, J. O. Dimmock, T. C. Harman, I. Melngailis, *Phys. Rev. Lett.* **23**, 7 (1969).
8. E. D. Hinkley, *M.I.T. Lincoln Lab. Solid State Res. Rep. 1968:3* (1968), p. 19; T. C. Harman, A. R. Calawa, I. Melngailis, J. O. Dimmock, *Appl. Phys. Lett.* **14**, 333 (1969).
9. C. K. N. Patel and E. D. Shaw, *Phys. Rev. Lett.* **24**, 451 (1970); A. Mooradian, S. R. J. Brueck, F. A. Blum, *Appl. Phys. Lett.* **17**, 481 (1970).

10. A. R. Calawa, T. C. Harman, M. Finn, P. Youtz, *Trans. Met. Soc. Amer. Inst. Mining Eng.* **242**, 374 (1968).
11. E. D. Hinkley, T. C. Harman, C. Freed, *Appl. Phys. Lett.* **13**, 49 (1968).
12. E. D. Hinkley, *ibid.* **16**, 351 (1970).
13. ——— and C. Freed, *Phys. Rev. Lett.* **23**, 277 (1969).
14. J. S. Garing, H. H. Nielsen, K. Narahari Rao, *J. Mol. Spectrosc.* **3**, 496 (1959).
15. H. Brunet and M. Perez, *ibid.* **29**, 472 (1969).
16. A. Turk, S. M. Edmonds, H. L. Mark, G. F. Collins, *Environ. Sci. Technol.* **2**, 44 (1968).
17. A spectrometer has been inserted in front of the detector for the direct absorption experiments in order to measure the laser wavelength and to serve as a band-pass filter permitting radiation in a single longitudinal mode to be transmitted, while excluding that which may be simultaneously present in adjacent modes. In a field instrument this refinement is not necessary.
18. M. C. Teich, *Proc. Inst. Elec. Electron. Eng.* **56**, 37 (1968).
19. R. H. Kingston, personal communication.
20. Although compositions with x approaching unity should produce lasers that emit near 3.7 micrometers at liquid helium temperatures, high carrier concentrations have, so far, thwarted attempts to produce such lasers (I. Melngailis, personal communication).
21. L. N. Kurbatov, A. D. Britov, I. S. Aver'yanov, V. E. Mashchenko, N. N. Mochalkin, A. I. Dirochka, *Sov. Phys. Semiconductors* **2**, 1008 (1969).
22. We gratefully acknowledge the many technical contributions and continued encouragement of R. H. Kingston. Appreciation is expressed to T. C. Harman and A. R. Calawa for providing the $Pb_{1-x}Sn_xTe$ material and many of the diode lasers used in this work. We thank J. O. Sample and F. H. Caswell for laser fabrication and general technical assistance. Work supported by the Air Pollution Control Office of the Environmental Protection Agency. The tunable diode laser was developed with the support of the Department of the Air Force and the Advanced Research Projects Agency of the Department of Defense.

Control of Particulate Air Pollutants:
Technical Aspects

Comparison of Filter Media Against Liquid and Solid Aerosols

RONALD G. STAFFORD and HARRY J. ETTINGER

Several filter media were evaluated by using liquid (0.3-μm DOP) and solid (approximately 0.3-μm polystyrene latex) test aerosols. Data were obtained for each medium at velocities ranging from 4 to 190 fpm. Below 100 fpm, Whatman 41 filter paper has a significantly higher efficiency when tested with solid aerosols. Below 40 fpm, IPC 1478 also has a significantly higher efficiency when tested with solid aerosols. Instrumentation limitations prevented measuring any significant differences for the high-efficiency media tested. Aerosol agglomeration and charge effects were eliminated by using an impactor and by establishing charge equilibrium for the solid aerosols. Average electron charge on latex particles was reduced to 0.7, and the test aerosol consisted of 86 to 97% single particles. The effects of filter loading on efficiency and pressure drop was also measured. Whatman 41 efficiency was significantly increased owing to loading with polystyrene latex aerosols. This did not occur with DOP.

Background and Introduction

CONSIDERABLE EFFORT has been directed toward defining efficiency of fibrous filters used for air cleaning and sampling. Minimum acceptable efficiency is often rigidly defined to minimize atmospheric contamination or to accurately estimate potential health hazards. These criteria frequently specify use of a solid or liquid aerosol of a specific size or size distribution.[1-4] In the United States, DOP (dioctyl phthalate) is the test aerosol most frequently used, and a considerable amount of filter efficiency data has been developed with this aerosol.[5-7] In addition, some data have been published defining efficiency characteristics using solid test aerosols.[8-13] Where similar media were tested

This work was performed under the auspices of the U. S. Atomic Energy Commission.

with aerosols of approximately the same size, comparison of test results show that a higher efficiency was measured when the solid aerosols were employed. Filtration theory does not indicate that these differences should exist. Differences have generally been attributed to (1) more agglomeration of the solid aerosol as opposed to the liquid aerosol; (2) electrical charging effects on the solid aerosol, which is minimal with a liquid aerosal; (3) increased and more rapid filter loading with a solid aerosol; (4) shatter of the liquid aerosol during collection on the filter fiber; (5) transport of the liquid aerosol during collection; or (6) small differences in the size and density characteristics of the various aerosols. Since most filters are tested with liquid aerosols, and in practice are required to collect solid aerosols, we felt that this discrepancy should be resolved. The high variability in test results and incomplete detailing of experimental procedures reported by several investigators using 0.3-μm DOP against Whatman 41 filter paper[5,6] indicate that it is impossible to reliably compare test data reported by different investigators using different aerosols.

A test program was therefore initiated to measure the efficiency of several common filter media using liquid and solid aerosols at different filtration velocities. DOP was selected as the liquid aerosol, since it is the most common aerosol used for filter efficiency tests in the United States.[1] Polystyrene latex as supplied by Dow Chemical Co. was selected as the solid aerosol because it is available in the 0.3-μm size range, it is monodispersed and spherical, and it has a density close to that of DOP.

Theory

Theoretical work indicates that a submicron particle will exhibit minimum collection efficiency through a fibrous filter.[14-16] This size has been calculated to be approximately 0.1 to 0.4 μm, and some experimental investigators have verified these theoretical

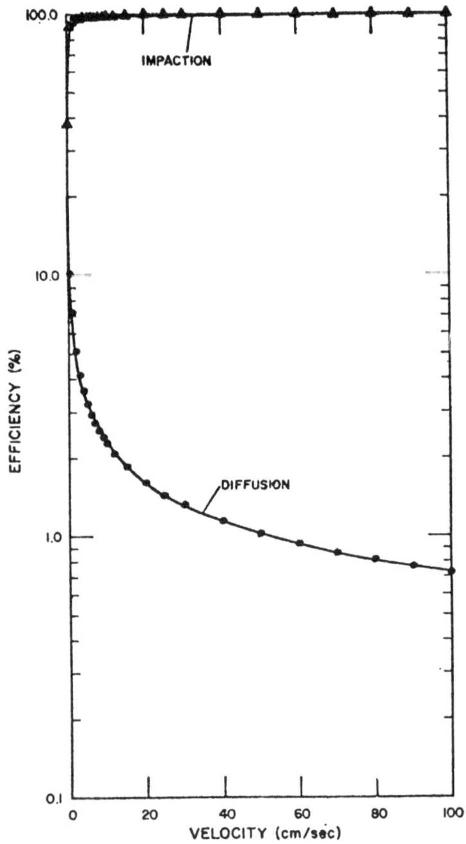

FIGURE 1. Single-fiber theoretical efficiencies due to impaction and diffusion using 0.312-μm polystyrene latex.

calculations,[8,11,17,18] with some discrepancy in defining the exact size for minimum filter efficiency. Other experimental investigators have found that efficiency continually decreases with decreasing particle size.[9,15,19]

Additional criteria which are important in filter efficiency characteristics are air velocity through the filter media, filter charge, filter fiber diameter, filter pressure drop, and filter loading characteristics. This study was not concerned with variations in particle size but compared efficiencies for liquid (0.3-μm DOP) and solid (0.234-, 0.264-, and 0.312-μm polystyrene latex) aerosols at various velocities. For a given particle size, theory predicts that there will be a velocity allowing minimum filter efficiency.[14,15,20] To provide an anticipated estimate of efficiency, the effect of the various theoretical single-fiber filtering mechanisms were calculated. For 0.3-μm aerosol particles and Whatman 41 filter media, direct interception and gravitational effects were calculated and were found to be negligible. The effect of electrical charge on efficiency was also considered. Whitby and Liu[21] have calculated that the average elemental charge on a 0.3-μm solid aerosol particle in Boltzmann equilibrium with a bipolar ion atmosphere will be 1.247; therefore, the theoretical aerosol charge was also considered to be negligible.

Collection efficiency attributed to diffusion was calculated from the diffusion coefficient formula,[22]

$$D_{BM} = \frac{CkT}{3\pi\mu d_p}$$

and the single-fiber diffusion collection efficiency formula,[23]

$$\eta(\mathrm{DIFFUSION}) = \left(\frac{8\,D_{BM}}{V d_f}\right)^{1/2}$$

Single-fiber efficiency due to inertial impaction was estimated by the following formula:

$$\eta(\text{impaction}) = 1.00 - \eta(\text{diffusion})$$

These data are plotted in Figure 1, showing that diffusion is a significant parameter only at velocities below 10 fpm. Inertial impaction accounts for the majority of collection above this velocity, and the curve shows the minimum efficiency to occur at a velocity well below 1 fpm.

Total filter mat efficiency due to diffusion, interception, and impaction was calculated for Whatman 41 and IPC 1478 from equations developed by Davies.[24] This is discussed together with our experimental results.

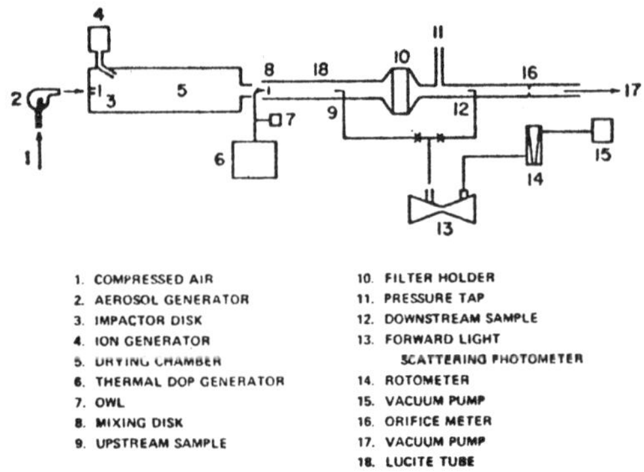

FIGURE 2. Schematic of filter test system.

Equipment

Polystyrene latex and DOP were selected as the solid and liquid test aerosols, respectively, because of their monodispersity and similar densities.

Figure 2 is a schematic of the test apparatus. A 10% dilution of polystyrene latex microspheres as obtained from Dow Chemical Co. was diluted to 0.25% with double-distilled water. This dilution corresponds to 1.57×10^{11} particles/ml of solution for 0.312-μm polystyrene latex. The latex suspension was aerosolized with a Penisol (Hudson Oxygen Therapy Sales Company, Los Angeles, California) nebulizer (2) operated at 15 psig. The aerosol was introduced into a drying chamber (5) through a 1/16-inch orifice and impacted on a 1-inch-diameter disk (3) to break up agglomerates produced by the nebulizer. An ion generator[25] (4) introduced a bipolar ion atmosphere which produced a charge equilibrium on the latex aerosol within the drying chamber. The aerosol then entered the test system consisting of a Lucite tube (18) with a mixing disk (8) to ensure proper mixing, a 2⅝-inch-diameter filter holder (10), an orifice meter (16) to monitor the airflow, and a vacuum source. (17).

DOP was generated from a Q-127 res-

TABLE I

Whatman 41 Filter Efficiency for Polystyrene Latex and DOP

Velocity (fpm)	0.312-μm Polystyrene Latex	0.264-μm Polystyrene Latex	0.3-μm Dioctyl Phthalate
4.0	66.80	58.83	55.98
6.0	68.32	61.18	57.43
9.0	69.14	54.85	59.26
13.0	72.54	52.50	59.00
17.0	72.46	59.38	58.76
18.0	70.69	59.38	60.71
20.1	75.99	62.72	66.00
22.4	—	65.69	63.65
24.4	—	65.79	68.92
26.0	80.86	65.39	72.76
29.8	—	71.93	70.93
32.8	87.18	81.88	74.80
35.5	—	76.08	73.29
38.1	88.02	82.37	77.44
42.6	92.04	79.10	77.26
45.6	—	82.39	78.46
46.4	93.32	80.47	79.71
50.2	93.72	86.39	83.79
53.5	95.72	85.35	82.28
56.6	95.08	89.52	84.39
59.5	95.81	90.44	85.12
62.2	95.51	90.89	85.28
64.6	96.25	91.27	85.81
66.6	96.15	93.38	88.38
69.2	95.71	92.10	87.59
71.1	96.64	93.77	88.62
74.8	97.07	92.30	90.36
78.1	95.44	92.31	90.54
82.6	96.73	92.53	90.43
85.5	96.58	92.16	91.45
88.4	97.29	94.86	92.36
95.1	97.42	93.46	93.29
100.8	97.48	96.00	93.77
106.5	97.93	97.37	94.79
111.2	97.85	97.52	95.06
116.0	98.00	95.70	93.77
120.7	97.22	97.20	95.48
125.6	97.84	95.69	95.91
129.2	97.90	94.11	96.12
133.8	97.66	96.67	96.37
137.7	98.29	97.49	96.85
140.3	97.70	98.39	97.44
144.1	97.94	98.48	97.54
160.0	97.52	98.81	98.16
171.9	98.05	—	99.02
183.0	98.16	—	99.38

pirator filter canister tester[1] (6). This unit heats DOP to 172°C and condenses the vapor by mixing dilution air to produce a monodisperse aerosol. Particle size was maintained at 0.3 μm by controlling the temperature of the diluting air. DOP size was monitored with an electronic "Owl" (7) which measures the polarization ratio of light scattered at 90° by the aerosol.[26] DOP size was also checked with a mechanical "Owl."[27]

Both upstream and downstream aerosol concentrations were measured with a forward light-scattering photometer (13) to determine

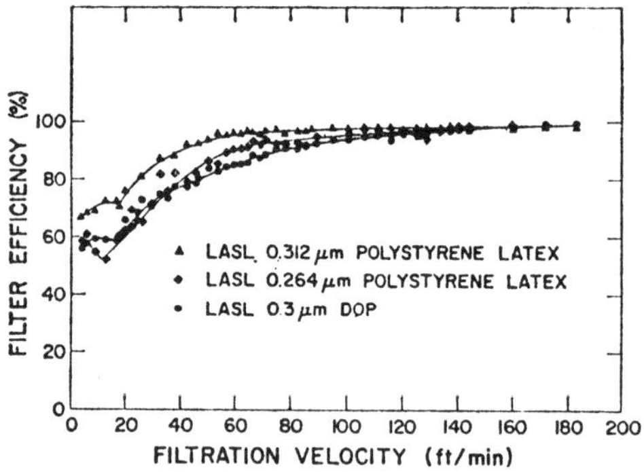

FIGURE 3. Filter efficiency versus filtration velocity for Whatman 41 filter media.

TABLE II
IPC 1478 Filter Efficiency for Polystyrene Latex and DOP

Velocity (fpm)	0.312-µm Polystyrene Latex	0.234-µm Polystyrene Latex	0.3-µm Dioctyl Phthalate
4	32.0	34.1	15.6
7	39.3	27.5	3.8
13	23.5	22.0	2.6
20	24.1	21.7	6.9
30	21.0	21.3	11.7
40	28.0	19.5	13.8
60	18.8	15.9	12.2
80	18.5	15.6	13.0
100	19.2	15.6	11.7
130	17.9	18.5	12.0
160	18.3	13.2	10.0
190	20.4	20.7	11.8

the filter efficiency. All readings were recorded on a Honeywell-Brown 12-inch strip chart recorder coupled to the photometer. A valve was incorporated on the upstream sampling port to produce a reduced pressure equal to the downstream sampling pressure. Aerosol loss in this valve was checked and found to be insignificant.

Aerosol agglomeration was monitored by collecting polystyrene latex directly on electron microscope grids using a point-to-plane electrostatic precipitator. Previous work has shown that this instrument defines representative aerosol size characteristics.[28] The number of single particles ranged from 86 to 97% as determined by electron microscopy. Doublets ranged from 14 to 2%, and triplets were

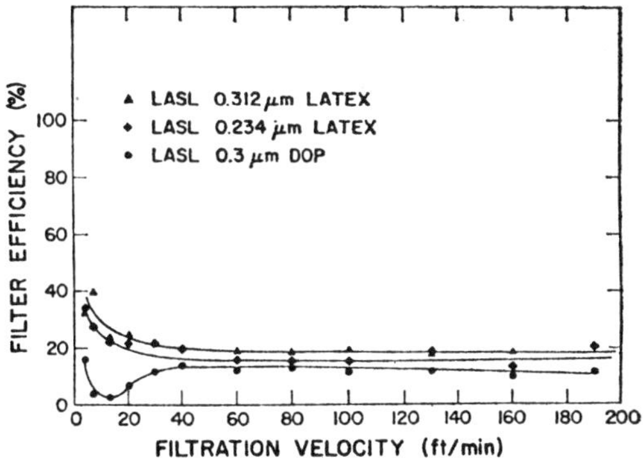

FIGURE 4. Filter efficiency versus filtration velocity for IPC 1478 filter media.

less than or equal to 1%. Assuming a binomial distribution, a chi-square test of independence was used to verify that a representative sample was examined. Three electron microscope grid samples were selected for each of the three polystyrene latex sizes. Eight photographs were taken of each grid. Using a two-way table for singlets and doublets, it was found that there was no difference between the samples at the 95% confidence level.

Latex aerosol charge was determined by using a modified Wesix Ion Collector together with a Keithley Model 640 vibrating capacitor electrometer which measured currents as low as 10^{-17} A. Modifications of the Wesix Ion Collector included lengthening of the plates which serve to polarize the assembly and collect the aerosol particles, decreasing the distance between them, and reducing the flow rate. These changes enabled us to measure the electric mobility (Z_p) as low as 5×10^{-5} cm²/volt-sec, which is the theoretical mobility for a 0.3-μm aerosol particle calculated from the equation[21]

$$Z_p = \frac{NeC}{3\pi d_p \mu}$$

Aerosol charge was found to be 0.7 electron charge per particle after neutralization. This is not significantly different from theoretical equilibrium conditions for a bipolar ion atmosphere which existed during the experimental tests.[21,29] Differences between measurement and theory may be due to small errors in generator output, dilution of latex suspension, impaction loss, or D. C. voltage change on the ion collector plates. The effect

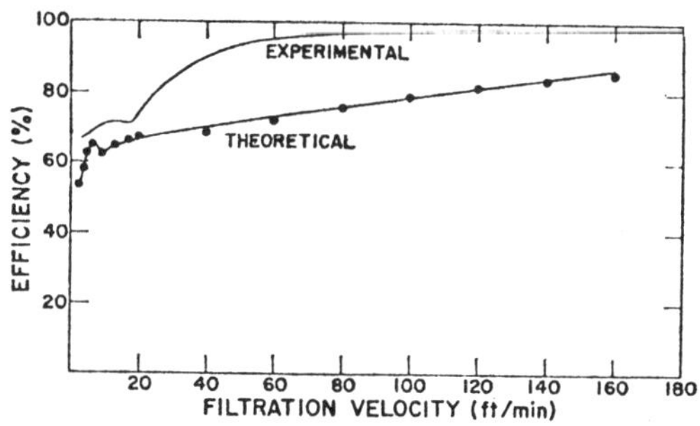

FIGURE 5. Whatman 41 total filter mat theoretical efficiency with 0.312-μm polystyrene latex.

of this small charge on filter efficiency was considered to be inconsequential.

Results and Discussion

Filter Efficiency Tests

Filter efficiency was measured for Whatman 41, IPC 1478, HV70, and CM 114 media with solid (0.312-, 0.264-, and 0.234-μm polystyrene latex) and liquid (0.3-μm DOP) aerosols. All efficiency data reported are averages of three values at each velocity. Table I and Figure 3 show the efficiency data for Whatman 41 at velocities from 4 to 183 fpm. Below 100 fpm, Whatman 41 exhibits significantly higher efficiencies with solid 0.312-μm polystyrene latex. Efficiencies with 0.264-μm latex are slightly less than those for 0.3-μm DOP at velocities below 40 fpm. Above 40 fpm, 0.264-μm latex produces higher efficiencies than 0.3-μm DOP.

IPC 1478 filter efficiency data are presented in Table II and Figure 4. Again, 0.312- and 0.234-μm polystyrene latex show a higher efficiency than 0.3-μm DOP over the entire

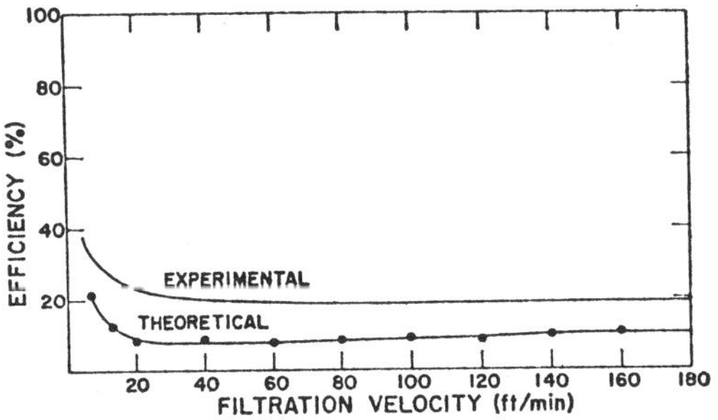

FIGURE 6. IPC 1478 total filter mat theoretical efficiency with 0.312-μm polystyrene latex.

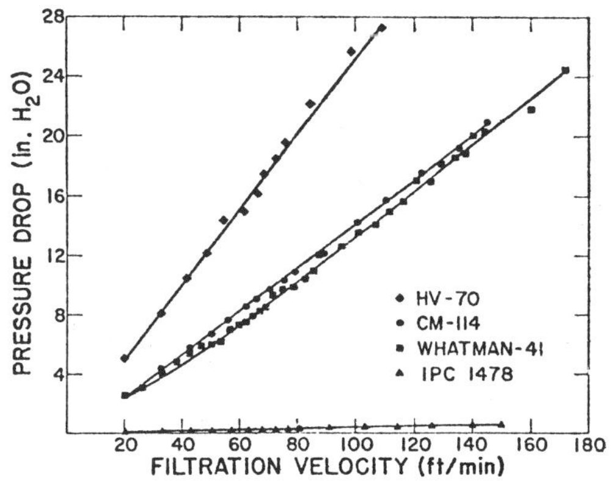

FIGURE 7. Filter pressure drop versus filtration velocity for different filter media.

velocity range tested. At velocities below 40 fpm this difference rapidly increases because latex does not exhibit the minimum efficiency present when DOP is used. The minimum efficiency occurs at a velocity slightly higher than single-fiber theory predicts. This may possibly be accounted for by differences in packing density of filters with the same fiber diameter. Owing to the high efficiencies of glass CM 114, and HV70, we were unable to detect any difference with the two types of aerosols. Our data for Whatman 41 and IPC 1478 are significantly higher than those reported by NRL[5] and HASL.[8] These differences may be accounted for by differences in quality control of the filter media as well as experimental techniques. Because charge and agglomeration were found to be insignificant in our studies, we must conclude that there is, in fact, a difference in efficiency of filter media with liquid and solid test aerosols.

Data are shown in Figures 5 and 6 comparing theoretical and experimental results. Experimental efficiencies with 0.312-μm polystyrene latex are higher than theoretical efficiencies for both Whatman 41 and IPC 1478 media. Experimental results and theoretical calculations show similar trends for both media.

Pressure Drop Tests

Figure 7 shows filter pressure drop at Los Alamos as a function of velocity for Whatman 41, IPC1478, HV70, and CM 114. Pressure drop of the test system was negligible for all velocities tested. Comparison shows that our pressure drop data are lower than those reported by NRL and HASL. The possibility that discrepancies were due to differences in atmospheric pressure was considered. Tests were conducted at increased pressures to simulate atmospheric pressure at sea level. These data are shown in Figure 8. At velocities below 50 fpm, it appears that multiplying by the ratio of the square root of the barometric pressure at sea level to that at Los Alamos would bring our pressure drop values closer to data obtained at sea level. Above 50 fpm there appear to be mechanisms occurring that cancel the effect of differences in pressure, which agrees with theory.[22] Quality control in the manufacture of these filter media may be an important factor in discrepancies between different laboratories.

Loading Tests

Simultaneous upstream and downstream aerosol concentrations were obtained for all four types of filter media with both polystyrene latex and DOP. Change in efficiency as a result of aerosol loading was not seen for any of the media with DOP as the test aerosol. Whatman 41 showed an increase in efficiency due to loading with polystyrene latex. Figure

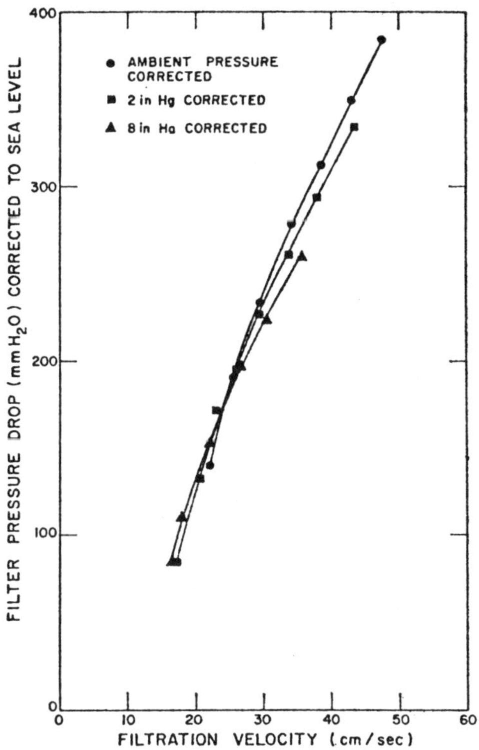

FIGURE 8. Filter pressure drop versus filtration velocity and increased pressure for CM 114 media.

9 shows the ratio of the efficiency at time t to efficiency at time t_0 as a function of time for Whatman 41 with 0.312-μm polystyrene latex. Because of the low concentration of latex aerosol (10^9 particles/min in air), we could not relate loading to an increase in weight of the media. Lindeken *et al.*[11] have shown a similar loading pattern for Whatman 41 with 0.365-μm polystyrene latex. IPC 1478 showed no change in efficiency after testing with polystyrene latex over a 30-minute test period. Instrumentation difficulties precluded measuring significant efficiency changes during aerosol loading for HV 70 and CM 114.

Summary

Whatman 41, IPC 1478, HV 70, and CM 114 filter media were used to compare liquid

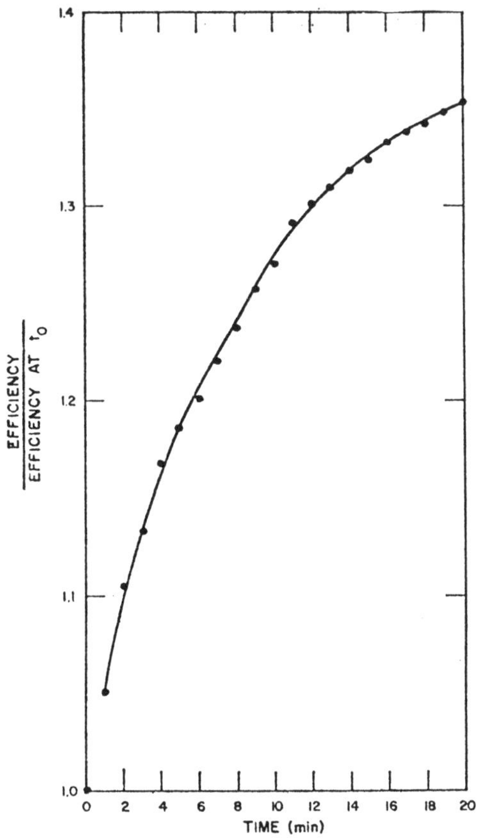

FIGURE 9. Filter loading for Whatman 41 filter media (filtration velocity = 40 fpm).

and solid test aerosols. Solid (0.312-, 0.264-, and 0.234-μm polystyrene latex) aerosols showed a higher efficiency than liquid (0.3-μm DOP) aerosols with Whatman 41 at velocities below 100 fpm. IPC 1478 also demonstrated a difference between the two types of aerosols; however, the variation was seen over the entire velocity range that was tested. Below 40 fpm the difference became much more pronounced. Because of the high efficiency of HV 70 and CM 114, instrumentation limitations prevented distinguishing differences between liquid and solid aerosols.

These tests indicate that there is a difference in filter efficiency for solid and liquid aerosols, and for more realistic tests a solid aerosol should be used.

Loading tests were performed on all four filter media with both polystyrene latex and DOP. No loading was observed when DOP was the test aerosol. Whatman 41 showed an increase in efficiency due to loading with polystyrene latex. IPC 1478 did not show any increase in efficiency with polystyrene latex over a 30-minute test period. Instrumentation limitations precluded measuring significant efficiency changes due to loading for HV 70 and CM 114.

Nomenclature

- C Cunningham correction factor (dimensionless)
- D_{BM} Diffusion coefficient (dimensionless)
- d_p Particle diameter (cm)
- d_f Filter fiber diameter (cm)
- e Elementary unit of charge (esu)
- k Boltzmann constant (ergs/°K)
- N Number of elementary units of charge (dimensionless)
- T Temperature (°K)
- V Upstream velocity of gas far removed from the effect of fiber (cm/sec)
- Z_p Mobility of aerosol particles (cm^2/volt-sec)
- η Filter efficiency (%)
- μ Coefficient of viscosity of gas (poise)

References

1. U. S. Army Chemical Corps and U. S. Navy Bureau of Ships: *Filter Units, Protective Clothing, Gas Mask Components and Related Products: Performance-Test Methods.* Report MIL-STD-282 (May 1956).
2. Parrish, E. C., and R. W. Schneider: *Tests of High Efficiency Filters and Filter Installations at ORNL.* Report ORNL-3442 (May 17, 1963).
3. Bureau of Mines: *Respiratory Protective Apparatus, Tests for Permissibility, Fees: Sched. 21B, Part 14, Filter-Type Dust, Fume, and Mist Respirators* (January 19, 1965).
4. Dill, R. S.: A Test Method for Air Filters. *ASHVE Trans. 44:* 379 (1938).
5. Lockhart, L. B., Jr., and R. L. Patterson, Jr.: *Characteristics of Air Filter Media Used for Monitoring Airborne Radioactivity.* NRL Report 6054 (March 20, 1964).

6. SMITH, W. J. and N. F. SUPRENANT: Properties of Various Filter Media for Asmospheric Dust Sampling. *Amer. Soc. Testing Mater. Proc. 53:* 1122 (1953). (Data reproduced in *Filters, Sorbents, and Air Cleaning Systems as Engineered Safeguard in Nuclear Installations,* ORNL-NSIC-13. October 1966).

7. ETTINGER, H. J., J. D. DEFIELD, D. A. BEVIS, and R. N. MITCHELL: HEPA Filter Efficiencies Using Thermal and Air-Jet Generated Dioctyl Phthalate. *Amer. Ind. Assoc. J. 30:* 20 (1969).

8. RIMBERG, D.:Penetration of IPC-1478, Whatman 41, and Type 5G Filter Papers as a Function of Particle Size and Velocity. *Amer. Ind. Assoc. J. 30:* 394 (1969).

9. STERN, S. C., H. W. ZELLER, and A. I. SCHEKMAN: The Aerosol Efficiency and Pressure Drop of a Fibrous Filter at Reduced Pressures. *J. Colloid Sci. 15:* 546 (1960).

10. POSNER, S.: Air Sampling Filter Retention Studies Using Solid Particles. *Proceedings of the 7th AEC Air Cleaning Conference.* TID-7627, p. 43 (October 1961).

11. LINDEKEN, C. L., R. L. MORGIN, and K. F. PETROCK: Collection Efficiency of Whatman 41 Filter Paper for Submicron Aerosols. *Health Phys. 9:* 305 (1963).

12. ADLEY, F. E., R. H. SCOTT, and W. E. GILL: *A Study of Efficiencies and Pressure Drop Characteristics of Air Filtering Media.* HW 28065 (August 1953).

13. ETTINGER, H. J., T. M. PANOWSKI, and C. E. BRECKINRIDGE: *Efficiency of Whatman 41 Filter Media against a Polydisperse Solid Aerosol: Interim Report.* LA-3898-MS (June 1968).

14. LANGMUIR, I., W. H. RODEBUSH, and V. K. LAMER: *Filtration of Aerosols and the Development of Filter Materials.* OSRD 865 (September 1942).

15. RAMSKILL, E. A., and W. L. ANDERSON: The Inertial Mechanism in the Mechanical Filtration of Aerosols. *J. Colloid Sci. 6:* 416 (1951).

16. GILLESPIE, T.: The Role of Electric Forces in the Filtration of Aerosols by Fiber Filters. *J. Colloid Sci. 10:* 299 (1955).

17. THOMAS, J. W., and R. E. YODER: Aerosol Size for Maximum Penetration through Fiberglas and Sand Filters. *A.M.A. Arch. Ind. Health 13:* 545 (1956).

18. DYMENT, J.: *The Penetration of Fibrous Media by Aerosols as a Function of Particle Size.* Conf. -337-51 (August 1963).

19. LAMER, V. K.: *Studies of Filtration of Monodisperse Aerosols.* NYO-512 (March 31, 1951).

20. DORMAN, R. G.: The Role of Diffusion, Interception and Inertia in the Filtration of Airborne Particles. In *Aerodynamic Capture of Particles* (E. G. Richardson, ed.), pp. 112-122, Pergamon Press, New York, London, Oxford, and Paris (1960).

21. WHITBY, K. T., and B. Y. LIU: The Electrical Behavior of Aerosols. In *Aerosol Science,* (C. N. Davies, ed.), pp. 59–86, Academic Press, London and New York (1966).

22. CHEN, C. Y.: Filtration of Aerosols by Fibrous Media. *Chem. Rev. 55:* No. 3 (June 1955).

23. STAIRMAND, C. J.: *Trans. Inst. Chem. Engrs. (London) 28:* 130 (1950).

24. DAVIES, C. N.: The Separation of Airborne Dust and Particles. *Inst. Mech. Eng. Proc.* (B) LB 203 (1952).

25. WHITBY, K. T.: Generator for Producing High Concentrations of Small Ions. *Rev. Sci. Instr. 32:* 1351 (1961)

26. U. S. Atomic Energy Commission: *Handbook on Aerosols,* Chapter 7, Washington, D. C. (1950).

27. LAMER, V. K., and D. SINCLAIR: *Measurement of Particle Size in Smoke, the Owl.* OSRD-1668 (August 24, 1943).

28. ETTINGER, H. J., and S. POSNER: Evaluation of Particle Sizing and Aerosol Sampling Techniques. *Amer. Ind. Hyg. Assoc. J. 26:* 17 (1965).

29. KEEFE, D., P. J. NOLAN, and T. RICH: *Proc. Roy. Irish Acad.* 27-45 (1959).

Particulates in Domestic Premises

I. Ambient Levels and Central Air Filtration

Neville M. Lefcoe, MD, FRCPC, and Ion I. Inculet, MEEng

Air particulates were surveyed in a home with central ventilation and an electrostatic filter in the main air duct with the returning air sampled. The average counts, in the 0.3μ, 0.5μ, and 1.0μ ranges, were related to the normal activities, children playing, house cleaning, and smoking, with the following observations: Counts in all ranges under minimal activity were significantly different when the filter was on compared to when the filter was off. Cleaning and dusting in the home overwhelmed the filter. Smoking one cigar in the home raised particle counts from 10 to 100 times. These particle counts stayed up for at least three hours when the filter was off. The counts returned towards base line levels in one to two hours with the filter on.

THE RELATIONSHIP of the symptoms of allergic respiratory tract disease, asthma and hay fever, to particulate air contamination has been well known for many years. House dust has been implicated in atopic individuals especially as a major factor, as shown by inhalation challenge tests and skin testing. The therapeutic practice of attempting to separate patients from undue exposure to house dust is widespread among physicians.

Recent investigation has attempted to identify the materials in house dust which may be responsible for the allergic response.[1,2] However, the nonspecific effect of particulates has also received a good deal of attention, and it is apparent that normals, as well as asthmatics and chronic bronchitics, respond by bronchoconstriction to challenges with inhaled particulates of various sizes.[3]

Definitive clinical work has been done in the area of seasonal pollenosis, and Friedlander and Friedlander[4] have observed that clinical symptomatology generally improved within a half an hour after a pollenosis subject entered on air-conditioned environment. Similarly, Criep and Green[5] observed relief from symptoms of hay fever and pollen asthma with the patients in a room where air cleaning equipment was used. Spiegelman and Friedman[6] showed that central air cleaning in the home would definitely change pollen counts, as well as bacterial counts, as compared to control observations without such filtration during the same season.

Previous work on house dust and its modification has dealt with an overall nonspecific measurement, such as dustfall (weight), and stain (coefficient of haze).[7]

The present study is an attempt to document the particulate load, in the smaller size ranges, in a modern home, equipped with central air filtering, under varying occupancy and activity conditions, and to compare particle counts of various sizes with and without filtration in the same home. Quantitation of the number of particulates in various size ranges, especially in the submicron range, has been made possible by the use of a light-scattering particle counter. No attempt was made in this experiment to count or identify the smaller number of larger particles.

The home selected is representative of thousands of such homes in this city, with standard furnishings and forced air heating, with the added feature of a central electrostatic air cleaner.

Materials and Methods

A modern split-level home with five bedrooms of 20,000 cu ft was used as the test house. A family of six, including four small children, were the normal occupants of the home, but most of the tests were done while the children, who returned only for lunch, were at school.

The central electrostatic filter was a standard commercial product, rated at 90% efficiency at 1,000 cu ft/min. The furnace fan was left on continuously, with a measured output on two occasions of approximately 1,000 cu ft/min, giving a total of three air changes per hour.

The particle counts were done with a particle counter (Bausch & Lomb 40-1) every hour commencing at 10 AM and finishing at 4 PM. Three counts were done in each size range one after the other, sampling for one minute at a time. Mean values were used, but there was virtually no variance in the three successive counts. The size ranges monitored were 0.3μ and above, 0.5μ and above, 1μ and above, and 2μ and above. Rather than collect separate samples from different rooms in the house, a mixed sample was obtained from the return air duct just upstream from the filter (Fig 1). Particle counts are per cubic foot of air. The nature of the activity in the house was recorded at hourly intervals during the day.

Counts were done for nine consecutive weekdays, alternating filter-on and filter-off on successive days.

Results

The table represents the gross data on seven of the nine days, separating the filter-on from the filter-off days in each size range. The two days not recorded in this group were days in which a cleaning woman was in the home, resulting in uniformly higher particle counts. It was felt that exclusion of these two days from the other days was justified on grounds of qualitatively different conditions existing in the home. Figure 2 shows the mean particle counts for the morning hours before noon hour activity resulted in raised counts. The 2μ range was not used, since the actual counts in this range were very low—largely one or zero. The *left-hand bar* of each pair represents mean particle count for the four days when the filter was off and the *right-hand bar*, the mean of the three days when the filter was on. Differences within each pair were significant at the 99.9% level of confidence. Figure 3 shows the effect of the filter in clearing the high particle counts which occurred when the breadwinner was at home for lunch and smoked one cigar. At 4 PM on the days the filter was off, the counts were still higher than the morning levels. However, by 4 PM on the day the filter was on, the particle counts in all three size ranges were back down to the morning range. By 3 PM they were trending well downward.

Unfortunately, no readings were procured in this series before the family had risen and before morning activity, ie, dressing and cooking breakfast, occurred. The differences in morning levels in the 0.3μ range may not reflect accurately the quiet nighttime and early morning levels. It may well be that the activity of the early morning has produced significant particle count elevation which has not completely dropped off by 10 AM—the first observation. This is borne out by Fig 3 where the reading at 11 AM is considerably lower than that at 10 AM, suggesting that the particle counts are still dropping from early morning activity on the filter-on days.

Figure 4 shows the counts at each hour on the two days in which a cleaning woman was active in the home, vacuuming and dusting. There is one day of observation under each filtering condition, with no significant difference in the counts.

Figure 5 compares two days of comparable minor activity, with the filter not functioning. However, on the day the windows remained open, the particle counts persisted at a high level all afternoon. Simultaneous outside counts were not obtained.

Comment

Since the objectives were not to analyze critically the performance of any specific

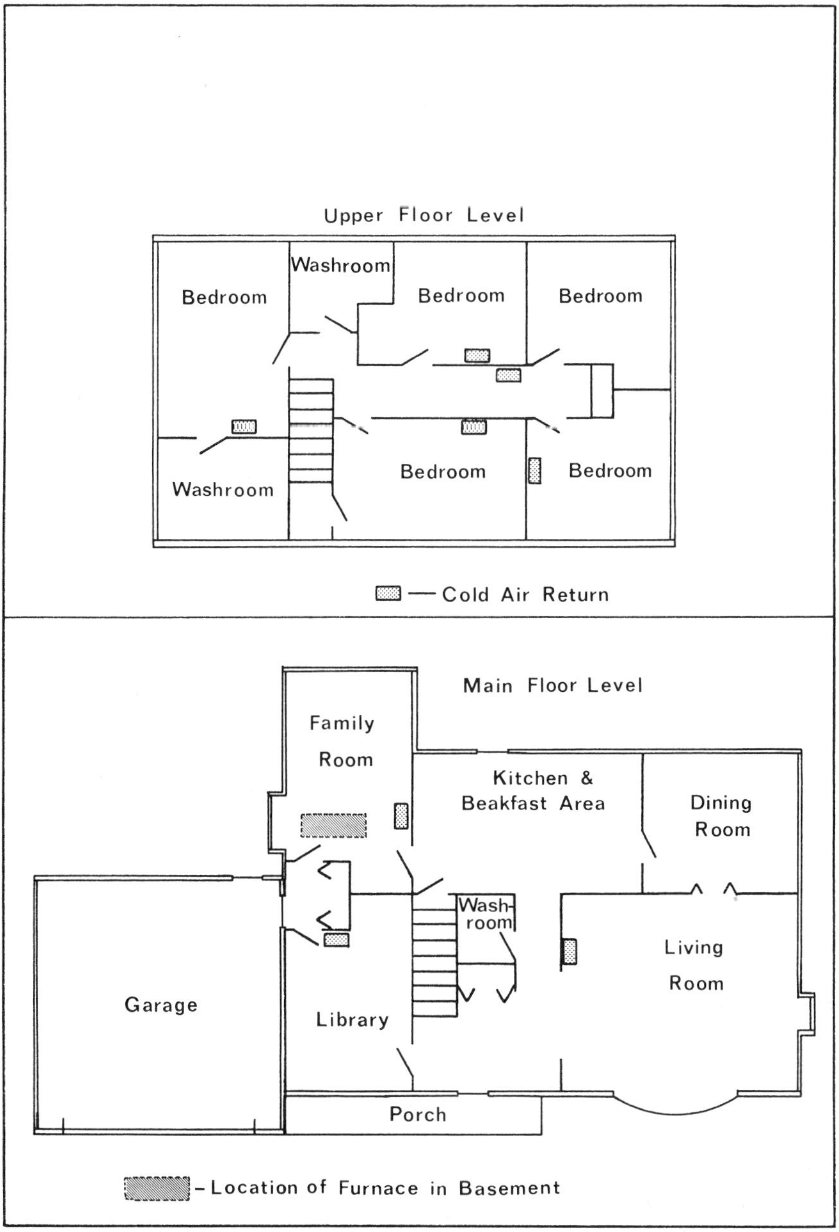

Fig 1.—Floor plan of test house showing return (cold) air ducts.

Particle Counts in 1,000's Per Cubic Foot of Air

		Particle Size							
		0.3μ		0.5μ		1μ		2μ	
		Electrostatic Filter							
Time	Date	On	Off	On	Off	On	Off	On	Off
10 AM	5/6	120	...	25	...	15	...	1.0	...
	5/7	...	1,000	...	350	...	180	...	1.5
	5/12	60	...	20	...	15	...	2.5	...
	5/9	...	1,000	...	250	...	150	...	1.0
	5/14	300	...	90	...	55	...	1.5	...
	5/13	...	350	...	70	...	35	...	1.5
	5/15	...	350	...	100	...	50	...	4.0
11 AM	5/6	100	...	20	...	12	...	0.5	...
	5/7	...	1,000	...	400	...	250	...	1.0
	5/12	12	...	3	...	3	...	0.5	...
	5/9	...	700	...	200	...	100	...	1.5
	5/14	180	...	40	...	25	...	1.0	...
	5/13	...	400	...	70	...	35	...	1.5
	5/15	...	400	...	90	...	50	...	1.5
12 NOON	5/6	100	...	25	...	20	...	1.0	...
	5/7	...	1,000	...	450	...	250	...	2.0
	5/12	30	...	18	...	12	...	1.0	...
	5/9	...	800	...	250	...	200	...	2.0
	5/14	130	...	35	...	18	...	1.0	...
	5/13	...	350	...	70	...	40	...	1.5
	5/15	...	350	...	80	...	50	...	2.5
1 PM	5/6	250	...	65	...	55	...	0.5	...
	5/7	...	1,000	...	500	...	300	...	1.5
	5/12	25	...	7	...	5	...	0.5	...
	5/9	...	600	...	250	...	150	...	12.0
	5/14	100	...	25	...	15	...	0.5	...
	5/13	...	500	...	130	...	80	...	2.0
	5/15	...	450	...	120	...	80	...	1.5
2 PM	5/6	650	...	150	...	70	...	1.0	...
	5/7	...	1,000	...	500	...	280	...	1.0
	5/12	1,000	...	800	...	450	...	1.5	...
	5/9	...	400	...	120	...	75	...	2.0
	5/14	600	...	200	...	90	...	0.5	...
	5/13	...	1,000	...	500	...	400	...	0.5
	5/15	...	450	...	120	...	70	...	2.0
3 PM	5/6	500	...	150	...	75	...	0.5	...
	5/7	...	1,000	...	450	...	250	...	1.5
	5/12	450	...	100	...	60	...	1.0	...
	5/9	...	350	...	100	...	70	...	1.5
	5/14	1,000	...	800	...	500	...	1.0	...
	5/13	...	1,000	...	500	...	300	...	0.5
	5/15	...	400	...	80	...	50	...	0.5
4 PM	5/6	500	...	150	...	80	...	0.5	...
	5/7	...	1,000	...	450	...	250	...	1.0
	5/12	80	...	20	...	12	...	0.5	...
	5/9	...	130	...	30	...	20	...	2.0
	5/14	500	...	150	...	80	...	0	...
	5/13	...	1,000	...	300	...	150	...	0.5
	5/15	...	1,000	...	350	...	200	...	1.5

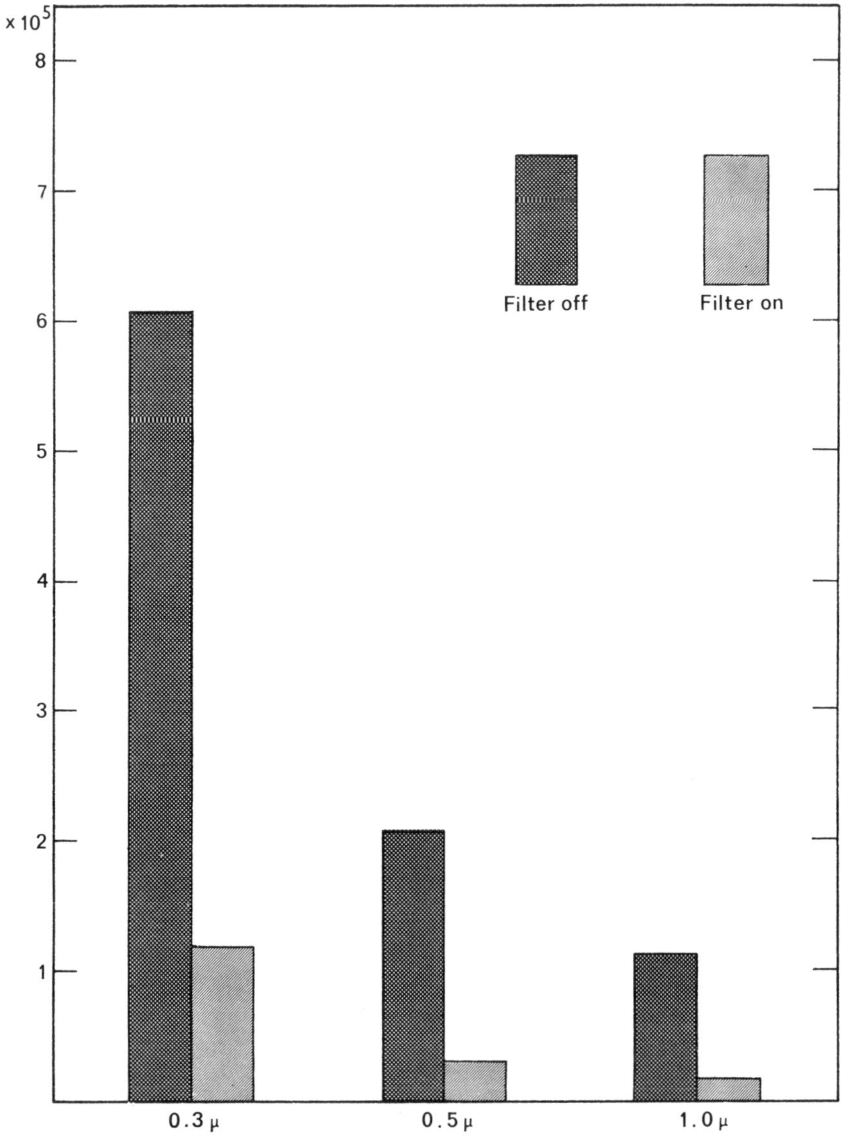

Difference in Means at 99.9% Level of Confidence (x 10^5):

4.92±2.93 1.77±1.62 0.96±0.90

Fig 2.—Comparison of mean morning counts particles per cubic foot filter-on vs filter-off.

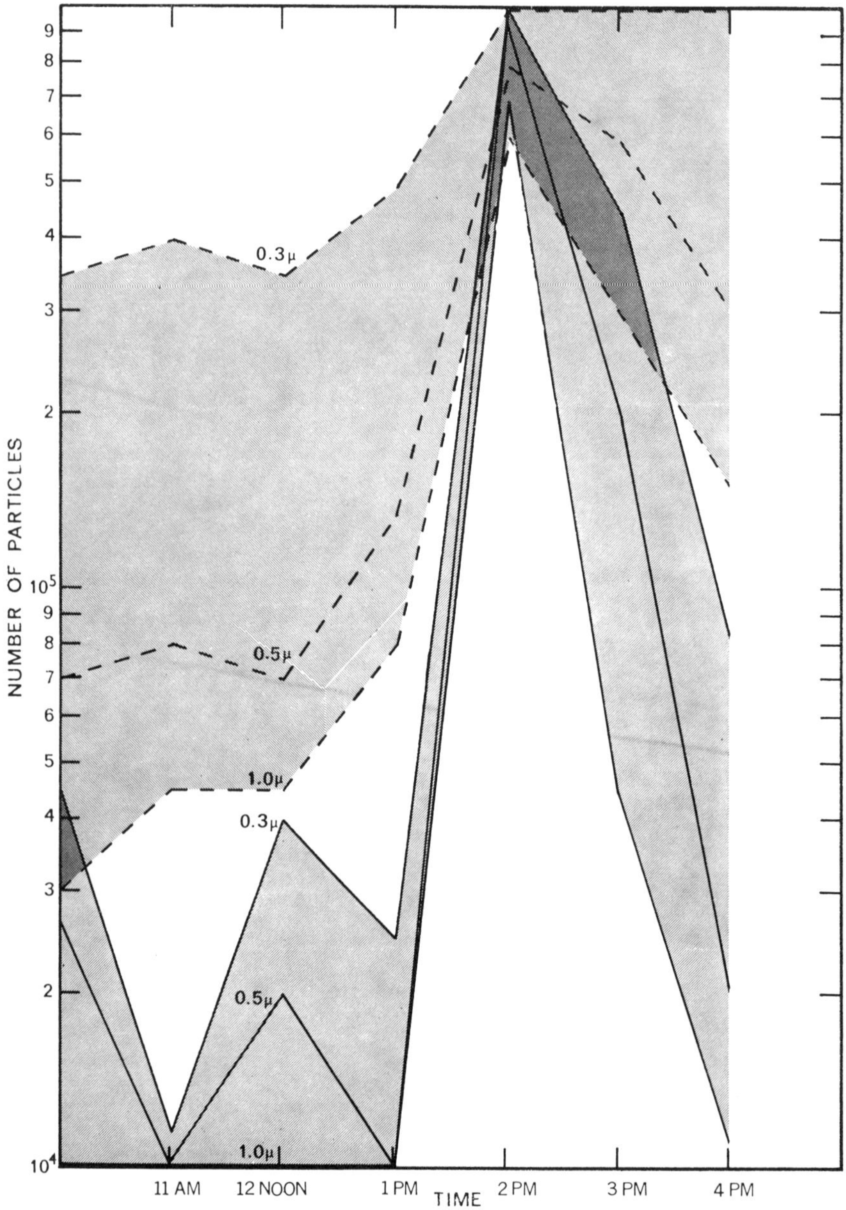

Fig 3.—Effect of smoking one cigar in living room between 1:15 and 1:45 PM. One day with filter on **(solid line)**, next day with filter off **(broken line)**.

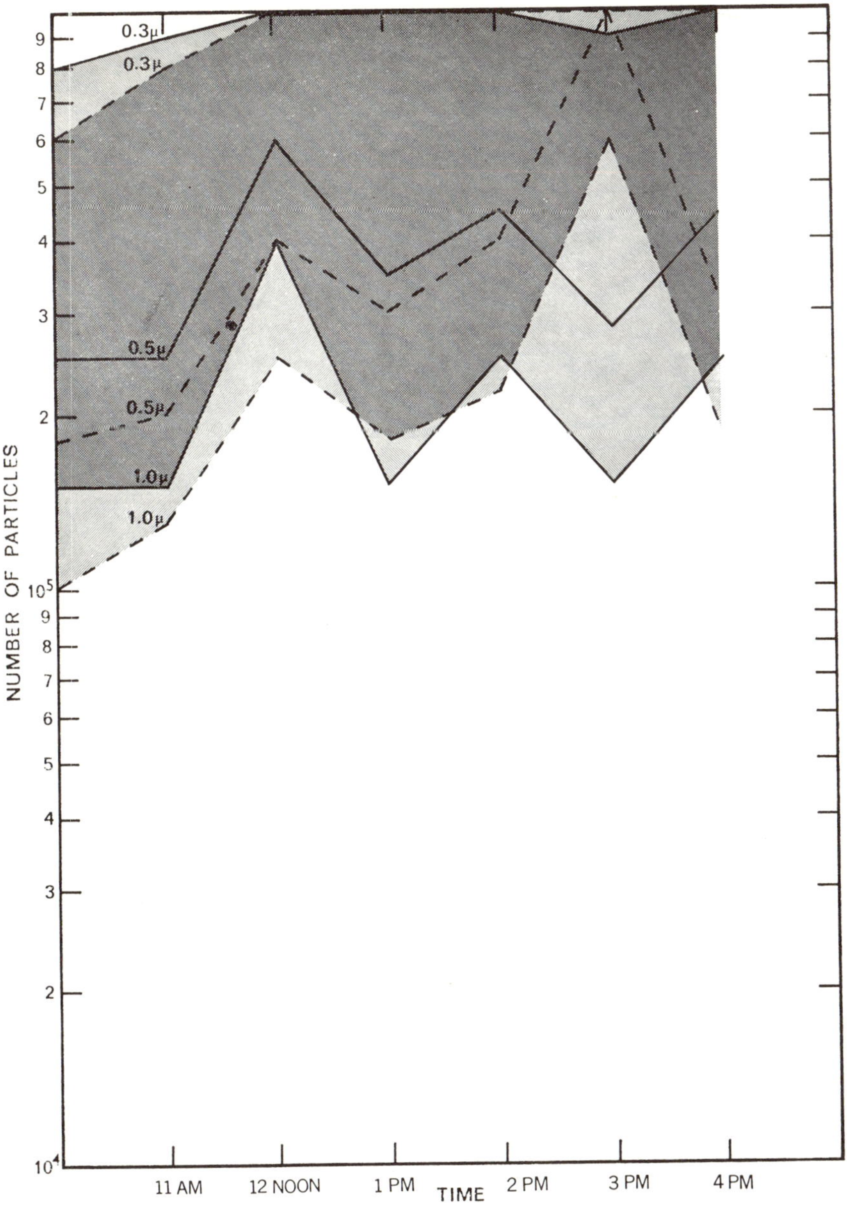

Fig 4.—Particle counts with cleaning woman working in the home, with no apparent reduction in counts on filter-on **(solid line)** day, compared with filter-off **(broken line)** day.

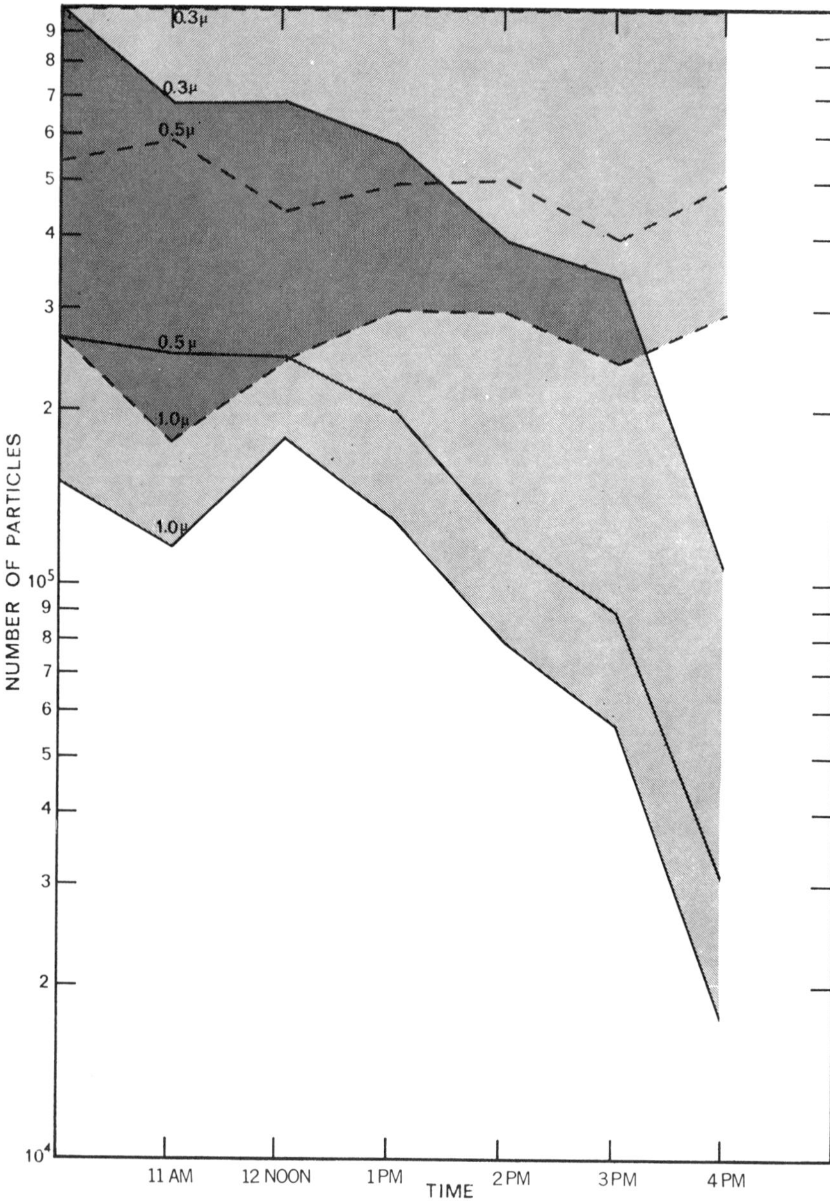

Fig 5.—Two days with filter off—comparing windows open (**broken line**) with windows closed (**solid line**).

commercial precipitator, no data are presented on the isolated precipitator; instead, the mean particle counts are shown in Fig 2, representing the dust loading in the total return air with the filter on and off. The comparisons in Fig 2 permit the calculation of the effective in situ working efficiency for each particle size. These effective efficiencies are as follows: 80% for the 0.3μ, 83% for the 0.5μ, and 86% for the 1μ particles.

The 10 AM readings of the level of particulate contamination in the home probably do not represent the resting (nighttime) level but rather a residual from early morning activity—7:30 AM to 9 AM, before the family left home.

There is evidence that smoking of one cigar completely overcomes the effect of the filtration device for at least one hour, and the counts are still well up at the end of two hours. The interesting thing is that in addition to being up in the 0.3μ and 0.5μ where presumably the vast majority of particulates from a cigar reside, there is also elevation in the 1μ micron range. Since no particle identification was attempted to this experiment, we are not entitled to say what they were. It is conceivable that this is not cigar smoke but other particulates produced by the activity in the home. Counts of this level were not produced even on the mornings when another adult was at home. They were, however, produced by dusting and cleaning.

On the days when a woman was vacuuming in the house most of the day, the particle count stayed up while she stopped for lunch and did not begin to diminish even at the conclusion of the experiment at 4 PM. The filter apparently had little effect on particle counts while this activity was going on. It can be argued that had we sampled in rooms in which the cleaning was not going on but in which the air had already passed through the filter and was mixing with air in the room, we might have noticed significant differences with the filter on as compared with the filter off. However, the normal activities in a home, closing and opening of doors, and air currents due to the forced air heating as well as natural convection, tend to produce a rapid mixing of the air from room to room. As the experiment was set up, there was continuous introduction, into the return air system, of dust from the room in which the cleaning was going on.

The moderately high level of counts with windows open and filter off was consistent. On comparable activity days also with filter off but with windows closed, the counts were lower, especially in the afternoon—providing there was no smoking in the house.

This work was supported by a grant from the Ontario Tuberculosis and Respiratory Disease Association.

The home of Mrs. A. Craig MacDonald was used in the study.

References

1. Voorhorst R, Spieksma-Boezeman MI, Spieksma FT: Is a mite (*Dermatophagoides sp.*) the producer of house dust allergens? *Allergie Asthma* **10**:329-334, 1964.

2. Mites and asthma, editorial. *Lancet* **1**:1295-1296, 1968.

3. Dubois AB, Dautrebande L: Acute effects of breathing inert dust particles and of carbachol aerosol on the mechanical characteristics of the lungs in man: Changes in response after inhaling sympathomimetic aerosols. *J Clin Invest* **37**:1746-1755, 1958.

4. Friedlander S, Friedlander AS: Effectiveness of a portable electrostatic precipitator in elimination of environmental allergens and control of allergic symptoms. *Ann Allergy* **12**:419-428, 1954.

5. Criep LH, Green MA: Air cleaning as an aid in the treatment of hay fever and bronchial asthma. *J Allergy* **7**:120-133, 1936.

6. Spiegelman MD, Friedman H: The effect of central air filtration and air conditioning on pollen and microbial contamination. *J Allergy* **42**:193-202, 1968.

7. Sutton DJ, Cloud HA, McNall PE Jr, et al: Performance and application of electronic air cleaners in occupied spaces. *ASHRAE J* **6**:55-88, 1964.

CONTROL OF PARTICULATE EMISSIONS FROM LIME PLANTS – A SURVEY

L. John Minnick

Lime has become the world's leading reagent for use in the treatment of both water and air pollution and, after sulfuric acid, is the No. 2 basic chemical in commercial use. As a result, the production capacities of the manufacturing plants are being rapidly expanded to meet the increasing demand for liming materials. This paper describes the achievements of the lime industry in developing methods of handling and controlling the various finely divided products which they produce. An extensive survey provides useful data on the availability and performance of many of the control devices that are currently in use, and an analysis is made of the operating efficiencies and costs of this equipment. The environmental control programs which are currently underway in this industry are described and an evaluation is made of these programs. The ultimate goals that are believed to be attainable are presented from the standpoint of emission control from individual processes as well as from operating plant complexes. While the paper deals primarily with practical operating and engineering aspects of the subject, some information is also included on methods of tests and the monitoring systems that are in use.

The manufacture of lime in America has been expanding during recent years to meet the growing demands of those industries involved with the control of air and water pollution. Lime slurries have been used for years in the control of stack gases particularly for the removal of SO_2. Currently there is interest in the injection of limestone into coal and oil fired boilers for the same purpose. In the field of water pollution control there are over 2000 municipalities in the United States who are using lime in their treatment plants. More than 50 types of industrial waste water are effectively controlled by the use of this reagent. These include such applications as steel waste pickle liquor, metal plating wastes, cannery wastes, wash water from chemical and pharmaceutical processes, laundries, petroleum refineries, etc. Recently several plants have been put on stream which are successfully removing phosphates from domestic water wastes with lime treatment.

The increase in demand for lime has required a substantial modification in the operations at many of the manufacturing plants. In view of the fine particle size of lime products, the industry has found it necessary to install equipment that can effectively handle huge tonnages of these fine powders. In addition, the control of particulate emissions has received a great deal of attention.

It is well known that the particulate emissions from a lime plant are not injurious to human health but they can be considered to be a "dust nuisance" to the community. A previous paper of the TI-2 Committee[1] describes the lime industry's problem of airborne dust in some detail and also provides some basis for the emission potentials of some of the equipment found in limestone processing, in lime burning, and lime hydration operations.

In order to provide technical information on the control of particulate emissions in lime plant operations, the National Lime Association Pure Air Committee has recently conducted an extensive survey of a number of lime manufacturing plants. The primary purpose of this paper is to provide the results of this survey together with some observations that have been made from the findings.

Unfortunately control of particulate emissions from lime plants is a complex subject, not only because of the diversity of the manufacturing processes used in the different plants, but because of the variations in control concepts that have been adopted by the plant engineers. A small lime plant which may not have rotary kilns can have entirely different emission problems than is the case with large operators who may have one or more rotary kilns. Some plants employ wet processing techniques whereas others use dry methods, and the newer plants may also be in a better position to utilize central system dust collection where this may be completely impractical for plants that have been established years ago.

Survey Findings

Twenty companies responded to a questionnaire circulated by the National Lime Association Pure Air Committee. The results of the survey are presented below in five sections, each of which represents a major segment of a lime plant operation.

In some areas the information available is quite limited; however, in other segments of the operation much more is known. It is expected that as supplementary information becomes available this report will be revised to keep the TI-2 Committee informed of significant progress in the lime industry's environmental control programs.

Quarrying

With proper planning of the quarry operation, this segment of lime production is not a major source of particulate emission, and effective control has been accomplished in many plants. Some of these procedures are outlined below.

Stripping Operations. The removal of the overburden, by the use of various earth moving equipment, usually results in large areas of bare or denuded soil exposed to the sun and wind erosion. The primary purpose of the control method is to retard the removal of moisture and to reduce the rapid movement of winds over the areas. This has been done by the construction of berms, by planting of suitable ground covers and windbreaks, erection of temporary fencing (snow fence), and the minimum disturbance of natural vegetation. Resulting spoil banks can be contoured and planted with suitable cover. The local county extension agents and state and federal conservation groups offer excellent advice and give valuable assistance to the operators in this respect. Examples of plants which perform well include crownvetch, multiflora rose, and hybrid poplars.

The final cleaning of the stripped or bare rock, exposed by the stripping operation, has been done by hand or mechanical sweeping of the surface. The development and use of a commercial mobile mounted vacuum pickup system has been suggested for dust control and final cleaning.

Primary Drilling. The prime source of dust from rotary and percussion drills using compressed air is closely confined in a small area. Various manufacturers have diverse selections of equipment to control effectively dust at this source. The most frequently used are mechanical cyclones. Bag type collectors are also available. Some states have made it mandatory to equip the drills with dust control devices. Figure 1 shows equipment used at one plant. Occasionally water with a wetting agent is effective.

Primary Blasting. The control of airborne contaminants from primary blasting is closely allied with good control of blasting practice. The initiation of detonation with multi-delay devices and the complete combustion of the explosive compounds during a well planned "shot" or "blast" results in low dust emission from primary blasting. The explosives manufacturers give excellent technical assistance for each operation's special problems.

Secondary Blasting or Breakage. The secondary blasting operation in many quarries is now either eliminated by better

Figure 1. Dust collection device used on quarry drill.

fragmentation during primary blasting, or by the use of "drop ball" cranes. Tractor mounted air or hydraulic operated "rock-splitters" have proven satisfactory for some operations and the use of these alternate methods has resulted in a safer, more economical, and cleaner operation. However, where secondary blasting is done, minimal generation of dust has been accomplished by good blasting practice.

Loading and Hauling. The major source of dust in these operations is the temporary haulage roads. The use of larger rubber tired haulage equipment can generate considerable dust on the crushed stone bed found on most roads. This problem is discussed in a later section dealing with fugitive dust.

Stone Processing

The processing of the stone obtained from the quarry includes such operations as crushing, pulverizing, screening, and conveying. In general, the control information obtained for these operations varied considerably. In nearly all plants water sprays were either an important adjunct of the control system, or were the sole control used.

Two companies made use of a centralized control system in which the dust emissions were collected at strategic points (bins, screens, crushers, transfer points) and led to high energy scrubbers having over 90% removal efficiencies. On the other extreme two large plants that purchase stone for calcining claimed that washing of the kiln feed has prevented dust problems in subsequent conveying and elevating.

Only two companies reported dust collection equipment (bag collectors) at the primary crusher. Water sprays are more common. For fine grinding, eight plants reported using bag collectors with no particular choice of cloth type. Collection points, in addition to the grinding mill, usually included associated feeders, elevators, screens, and belts.

Table I gives some details of these systems.

Lime Kilns

It is apparent from the response to the survey that dust control in lime kiln operations is a developing art since few specific trends can be detected and the methods of operation and results obtained cover a wide range. Some of the variance is undoubtedly due to imprecise measurement methods, design of equipment, and lack of experience in operating the control devices.

Nearly all of the industry's dust control problem on kilns pertains to the rotary kiln since this process accounts for 85% of all commercially produced quicklime. Vertical, Calcimatic, Fluo-Solids, and Corson kilns also produce particulate emissions, but there appears to be no problem in

Figure 2. Commercial vacuum sweeper.

suppressing the dust to allowable limits with baghouses, scrubbers, or cyclone-scrubber combinations. In the latter cases efficiencies are reported of 98.5% and above.

Most plants use primary collectors, consisting of dust chambers and/or cyclones, on the kilns with efficiencies ranging from 25 to 80%. These were followed by secondary collectors as follows:

Cyclones	20 kilns
Wet scrubbers	38 kilns
Baghouses	11 kilns
Electrical precipitator	1 kiln
No secondary collection	8 kilns

Some of the problems associated with the control of emissions from kilns have been reported as:

1. Disposal of dust collected. Most expeditious disposal of dry dust from baghouses is by screw conveyor from the hopper to a pneumatic conveying system to bin or silo. Ponding and lagooning of collected dust from wet scrubbers can be complicated. The settleable solids must be removed systematically by drag and loading into trucks for disposal. In addition, it is necessary to provide for disposal or reuse of the alkaline water.
2. Wear and tear on high capacity slurry pumping equipment.
3. Disappointingly short bag life of less than a year.
4. High abrasion and corrosion damage to fan impellers and dampers.

It was observed from the reports returned in the survey that both scrubbers and baghouses, if properly designed and maintained, have demonstrated high dust removal efficiency that should meet strict requirements. However, because of the high operating and maintenance costs for these, electrostatic precipitators have evolved new interest. Their higher capital costs may be more than offset by their inherently lower operating and maintenance costs. Finally, it was evident that no cyclonic method alone would be able to meet strict dust emission standards. A generalized estimate of costs for the various types of control equipment for a rotary kiln at two efficiency levels is included in Appendix I which also presents more detailed information on the control equipment on lime kilns.

Lime Hydration and Driers

The hydration of lime and hydrate bagging present a potential dust problem, although existing methods of control appear adequate in solving the problem. The main source of concern is the public reaction to the steam plume which cannot be readily eradicated.

The types of controls normally used in hydrators consist of the following:

1. Water sprays in hydrator stack. Dust particles are entrapped by sprays and the resulting slurry or milk-of-lime is piped back to the hydrator's pre-mixer as part of the slaking water. Lime losses are averted. Virtually all efficient hydrators have this equipment integrally installed. Wet cyclones or a small wet scrubber may be used instead or as a supplement to sprays. Most of what is vented is steam from the heat of hydration.
2. Cyclonic pre-cleaning may be performed at all transfer points in the hydration system, like the conveyor, elevator, quicklime grinder and screen, air classifier, milling, and bagging. This partially reduces dust load, but further dust reduction is necessary. Individual dry cyclones may also be employed at a specific location.
3. Emissions from a cyclonic pre-cleaner are usually vented into a baghouse or small independent bag filters may be

Table I. Stone processing.

Plant	Collection point[a]	Airflow	Cloth area sq ft	No. of bags	No. of compartments	Type cloth	Installed cost
1	Ball Mill	10,000 cfm at 80°F	1480	—	—	—	$19,000
2	Sec. Crushing and Screening	11,000 cfm at 14 in. H_2O	3600	64	4	Dacron	—
3	Raymond Mill	4000 cfm at 6–14 in. H_2O	1830	332	4	Glass	—
4	(2) Raymond Mills	7000 cfm at 9–17 in. H_2O	3000	240	3	Cotton	—
5	Pebble Mill	2000 cfm at 6 in. H_2O	1320	32	4	Glass	—
6	Pebble Mill	2000 cfm at 8 in. H_2O	2592	216	—	Sateen	—
7	Bradley Mill	6000 cfm	—	—	—	—	—
8	Raymond Mill	2000 cfm	—	—	—	—	—

[a] Including associated feeder, elevator, screens, and belts. One plant gave dust loading figures as 53.8 lb/hr at 150°F for the Raymond Mill; 11 lb/hr at 135°F for the tube mill; 48 lb/hr at 160°F at the conveyor.

operated at specific locations, like bag packers. Bag filters are often an integral part of packers. Hoods are also employed at baggers to confine dust.

One company that makes several types of hydrates employs six separate dust collection units: two wet cyclonic collectors connected with hydrator stack; bag filter for Raymond Mill; scrubber to control lime pumping to storage tank; tunnel collector; and a bag collector for quicklime grinding. Then, they used three bag filters in their three bagging machines.

Table II gives details of the statistical results obtained.

Driers represent, with the possible exception of rotary kilns, the most difficult dust control operation in a lime or limestone plant. For meeting strict emission standards it appears that primary cyclone collectors followed by baghouses as secondary are almost mandatory. Table III gives the figures obtained.

Fugitive Dust

The first legislative body to develop regulations which are specifically directed to fugitive dust as emitted from plant operations, such as the lime industry, is the State of Pennsylvania. The method of test which is required by this regulation is based on the use of a portable electrostatic precipitator.[2] In monitoring these emissions, measurements are made for ten minute periods downwind of the source at a location outside of the plant property. Average background concentrations are also determined upwind of the plant. The emission is calculated by subtracting the upwind from the downwind reading. The regulation also specifies a limit on the stack emissions based on ground level concentrations due to particulate fall (greater than 10 microns) and suspended particulate matter (less than 10 microns). In view of the increasing trend to establish limits based on the total emission from a facility rather than on separate sources such as stacks, the lime industry has also been making measurements of ambient air concentrations in the vicinity of the plants.

A review of the major sources of fugitive dust emission in lime plants is presented below with specific comments on various procedures that are utilized to reduce the contribution from these sources.

Curtailment of Dust from Roads. The effect of traffic and wind on dust which accumulates on roads is a primary source of fugitive dust. Since many of the roads in the plant complex are temporary, and may also be used at infrequent intervals, all of the roadways in the quarry and stone processing operations cannot be covered with a permanent wearing course. In those cases where roads can be paved, asphaltic mixtures have been applied over the base course either as an oil and stone chip covering, or as a plant mix asphaltic concrete wearing course. The cost of installations generally range from $2.00/ to $3.00/sq yd.

Table II. Hydrators

Type of system	Air flow cfm	Filter area, sq ft	Air-cloth ratio	No. of bags	No. of compart-ments	Capital and installed cost	Type of application
Bag filter	13,000	1750	—	128	—	19,000	Bagging
Bag filter	11,000 (5 in. w.g.)	9660	—	112	—	30,000	Fines from cyclone
Bag filter	—	1668	3.55:1	—	—	9,500	Fines from pre-cleaner
Bag filter	5,000	1050	—	—	—	—	—
Bag filter	10,000	—	—	—	3	30,000	Bagging
Bag filter	14,850	—	—	—	—	—	—
Bag filter	4,250	—	—	43	—	—	Bagging
Bag filter	6,000	576	—	—	—	—	Bagging
Bag filter	5,000	1050	—	—	—	10,000	Quicklime grinding
Bag filter	2,500 (150°F)	344	—	—	—	16,000	Raymond Mill
Scrubber	—	—	—	—	—	9,500	—
Cyclone pre-cleaner	—	—	—	—	—	10,000	—

Bags used are cotton, dacron, and glass; dacron most common.

Table III. Driers

Plant	Collecting system	Cloth area, sq ft	No. of bags	No. of compart- ments	Airflow, cfm	Efficiency	Capital and installed cost
1	Cyclone and baghouse	2550	360	—	15,000 at 220°F	—	$10,000 Cyclone 36,000 Baghouse
2	Baghouse	6250	112	4	4,500–7,500 at 12″	—	20,000
3	Primary, secondary cyclone, and baghouse	3120	312	8	18,100 at 230°F	99 + %	50,000
4	Cyclone (only)	—	—	—	8,800	(508 lb/hr)	12,000
5	2-Stage cyclone and high-energy scrubber	(400 gpm H_2O)	—	—	50,000 at 30″	99.9%	215,000

The removal of dust from the paved roads is carried out in several ways including the use of street cleaning equipment such as brush type and vacuum type sweepers. It has been the experience of some of the operators that the wet brush sweeper leaves a residual film of dust after the surface dries. A vacuum type sweeper, such as that manufactured by The Tennant Company or The Wayne Company, has been found to be quite satisfactory (Figure 2).

Dust control from unpaved roads is universally achieved by means of a periodic covering of oil or by chemical treatment. Some plants find that they can utilize waste crusher oil very effectively either as-is, or by diluting with small amounts of diesel fuel. The use of this particular method must take cognizance of other factors such as odor or leaching of the oil into adjacent streams. The rate of application of oil is in the neighborhood of 5 gallons per hundred square yards of roadway. The material is applied every one to two weeks depending on weather conditions. Where an asphalt base material is used (5–10% asphalt), the coverage is about half of that given above. This material need not be applied as frequently to the roadway. In addition, most plants apply water (with or without calcium chloride additive) as a supplement to the oiling program.

Stockpiles. An inherent part of lime plant production is the establishment of numerous surge stockpiles of limestone. This is primarily the result of accumulations of certain sizes of stone that may not be required by the market at the rate with which other sizes are utilized. Further, some of the crushed limestone sizes are supplied on a seasonal demand and the producer must therefore stockpile in advance. Fortunately most of these materials can be stockpiled in the open and may be kept in a moistened conditioned. As mentioned previously, waste piles which result from overburden or inclusions of impure stone can be a problem since there is usually very little demand for this type of material. Products such as quicklime or hydrated lime are always stored in sealed bins or silos and therefore do not constitute an "exposed" stockpile problem.

The industry has been reasonably successful in controlling the emission of airborne dust from stockpiles in several ways. The most important is the wetting down of the stockpiles on a continuing basis. This substantially reduces the effect of wind erosion. A photograph (Figure 3) is presented showing the type of equipment that is being used successfully for this purpose. These are spray towers approximately 40 ft high that are equipped with Rainbird type spray nozzles capable of spraying water at the rate of 500 gpm. The unit shown in the picture can spread a spray of water in a continuous circle of a 200-ft radius. It requires but three passes to wet down the pile. For those materials which must be shipped in dry condition, the stockpiles are either kept within suitable enclosures, or are run through a drying process

Figure 3. Rainbird spray towers wetting stockpiles.

Figure 4. Closed pneumatic piping system for conveying of dry materials.

Figure 5. Blower system widely used for unloading trucks.

at the time of shipment. Permanent waste piles or spoil banks that have been covered with plantings are reported to give adequate protection against wind erosion.

Proper location of stockpiles can be helpful particularly if they are behind natural (or manufactured) barriers, such as a hill or slope of ground, a row of trees, wind baffles, etc. The active or working side of a stockpile is usually established by considering the prevailing wind direction as well as traffic flow to and from the pile. A few plants are experimenting with special dust suppression coatings that may prove to be of value in some cases. There seems to be little benefit in controlling size or slope of the piles where preventive measures have been employed.

Truck Transportation. Many lime plants have reduced in-plant traffic to a substantial degree in the last few years. This has been accomplished through the increased use of belt conveyors which not only eliminate the use of dump trucks, but frequently eliminate intermediate surge pile areas. For material such as hydrated lime, air slides are now used in many applications.

Transport by means of air pumps (such as Fuller Kenyon) is widely used in situations where the material is transported over considerable distance (Table IV).

Figure 4 shows a conveying system typical of those used in many of the plant operations.

It is not within the scope of this report to describe all of the latest developments in the design of the various systems of truck transport of material. Fortunately most of the designs of pumper trucks do significantly control the particulate emission by keeping the transfer of material completely enclosed. Venting of air in the transfer is accomplished by the use of bag type collectors on the receiving silos. Figure 5 is an example of pumper truck operations.

Dump trucks are still widely used, particularly for the transportation of coarse aggregate. Current practice utilizes covering of the material with tarpaulins or spraying with water. These procedures are usually directed to truck transportation out of the plant operation. Dump trucks create

Table IV: Pneumatic Conveying Information

	Pulverized limestone	Quicklime	Hydrated lime
Capacity (t/hr)	9–12	10–14	10 12
Ranges of horsepower	40–100	100–150	20–40
Typical pumping heights (ft)	40–60	55–75	45–65
Horizontal pumping distance (ft.)	500–600	550–650	300–400
Approximate installed costs	$30,000	$50,000	$12,000

a dust problem in-plant when they are used to transport fine sizes of dry material to such points as stockpiles, loading hoppers for rail transport, etc. The most successful control for these types of operation is to provide enclosures over the dumping area. These enclosures may require dust collection equipment to minimize particulate emissions.

A very important requirement for truck operators is the limit of speed in (and also out of) the plant premises. While the lime plant has no control of the driver once he leaves the plant, a strenuous effort has been applied to keep speed limits at a low level. Within the plant operations the limit is usually posted at 10 miles per hour.

Miscellaneous. In addition to the items covered in this report, fugitive dust can result from a variety of problems within the plant complex. Spillage from overloaded trucks, leaky bins, and the occasional breaking of bags are examples of intermittent problems that need to be considered. In general, the control of these devices is accomplished by the use of careful houskeeping procedures. Control of isolated situations is only effective where the plant has a good surveillance and educational program. The plants who are attempting to do this have indicated that such a program is similar to the plant "safety" program which can only be successful where a continuous on-going educational effort is vigorously promoted by the plant management.

Resume of Survey Findings

The survey has shown that many lime plants have been able to control emission effectively to a degree that should meet the demands of reasonable air quality standards. Some of the plants have also pioneered in the control of emission of "fugitive dust" from such sources as roadways, stockpiles, transport equipment, etc.

A large variety of equipment is in use throughout the industry to control emissions from the different process operations. The majority of this equipment is performing satisfactorily; however, the survey has indicated that it will be quite difficult to standardize on the effectiveness of many of the control methods. The manufacturing diversity existing between plants is a problem which must be carefully considered if uniform standards are to be developed. In spite of these problems it is believed that the impressive evidence of available handling, conveying, and dust control procedures and technology which have been outlined in the survey should provide a basis for preparing effective emission standards (or regulations) for this industry. The lime companies expect to develop their own voluntary regulation program and it is believed that with the continuing cooperation of governmental officials, the industry can control the environment in the vicinity of quarries and lime processing operations.

The National Lime Association Pure Air Committee is evaluating all of the available information and is currently preparing suitable guide lines to assist interested parties in formulating specific control procedures that can be directly applied to the production of limestone and lime products.

Acknowledgment

The information presented in this report represents the efforts of a substantial number of contributors who have indicated their preference for remaining anonymous. The author, however, wishes to express appreciation to all of them for their help.

Most of the survey findings were analyzed by Robert Boynton (Executive Director of the National Lime Association) who applied considerable time in evaluating the information which he received. In addition, Frank Willard (Regional Air Pollution Control Engineer of the Pennsylvania Department of Health) and Lax Jain (Environmental Engineer of the Foote Mineral Company) a member of the National Lime Association Pure Air Committee, both provided valuable assistance in preparing this report.

References

1. Lewis, C. J. and Crocker, B. B., The Lime Industry's Problem of Airborne Dust. TI-2 Report No. 10, *J. Air. Poll. Control Assoc.*, 19 (1), 31 (Jan. 1969).
2. Pennsylvania Department of Health. Regulation IV—Guide Lines to Control Air Pollution from Sources of Particulate or Gaseous Matter Emissions. Adopted March 15, 1966.

Appendix I. Rotary Kilns

Seventy-two kilns were reported ranging from 50 to 650 tpd capacity.

Stack heights ranged from 40 to 120 ft in height. One stack height was reported at 250 ft servicing three large-capacity kilns.

Gas volumes ranged from 12,400 to 90,000 scfm with a mean value of 38,000.

Stack temperatures ranged from 100° to 900°F.

Fuels used—60% pulverized coal, 28% natural gas, 6% combination of coal and gas, and 6% Bunker C fuel oil.

Baghouse filter areas ranged from 70,000 sq ft and 12 compartment systems to 750 sq ft single compartment systems.

Air to cloth ratios ranged from 2.5:1 to 1.7:1 with a mean ratio of 2.2:1.

Efficiencies of collection ranged as follows:

1. Cyclones—30–93.3% (depending on primary or secondary dust removal)
2. Scrubbers—93–99.5%
3. Baghouses—98–99.8%

Estimate of Rotary Kiln Control Costs*

Capacity —350 tons/day
Exhaust Gas Volume —100,000 cfm
Exhaust Gas Temperature—550°F
Power Cost —$.01/kwhr

Type of collector	Cost of unit	Operating cost per year	Maintenance cost per year
	For 99% Efficiency		
Baghouse	$150,000	$14,000	$8,000
Electrostatic precipitator	200,000	2,300	2,000
Scrubber	60,000	65,000	6,000
	For 97% Efficiency		
Baghouse	$150,000	$14,000	$8,000
Electrostatic precipitator	100,000	2,300	2,000
Scrubber	50,000	32,000	6,000

* The estimates are based on purchase of equipment during the last fifteen years and do not represent current capital costs, many of which have increased because of inflation factors. In addition, the estimates do not include total installation and start-up costs.

Control of Gaseous Air Pollutants: Technical Aspects

RELATING AIR POLLUTANT EFFECTS TO CONCENTRATION AND CONTROL

Ralph I. Larsen

Examples are cited relating air pollutant effects to concentration and control. The examples are organized under a dozen possible steps that could be used to proceed from air quality criteria to clean air. It is hoped that this paper will be of some assistance to State, regional, and local air managers and their air management boards in deriving the air quality standards and emission standards needed to prevent various unwanted pollutant effects.

Publication of the first air quality criteria documents by the National Air Pollution Control Administration (NAPCA) in 1969 initiated several activities as specified in the Clean Air Act. These activities are discussed in a NAPCA publication, *Guidelines for the Development of Air Quality Standards and Implementation Plans.*[1] The "guidelines" have been prepared to assist the States in developing standards and implementation control plans.

For the purposes of discussion, the process of proceeding from air quality criteria to clean air, as discussed in the "guidelines," might be divided into about a dozen steps (listed below). This paper will concentrate on examples relating to the first five of these steps. Various types of air pollutant effects and their relationship to ambient air pollutant concentrations and control calculations will be discussed. It is not the purpose of this paper to recommend any particular numbers. It is merely hoped that the techniques discussed will be of some assistance to State, regional, and local air managers and their air management boards as they use the "guidelines," air quality criteria documents, and other information in deriving the numbers needed for their particular areas.

The process of proceeding from criteria to clean air, which might be termed air management, might be divided into the following steps:

1. Determine the effects of various pollutant concentrations and exposure durations on people, animals, plants, and property.
2. Decide which effects to prevent.
3. Select ambient air quality standards that will prevent these effects.
4. Measure and evaluate ambient air pollutant concentrations.
5. Calculate over-all source reductions needed to achieve selected ambient air quality standards.
6. Measure or estimate emissions from each source type in an area.
7. Decide how much each source type could be permitted to emit and still achieve the desired ambient air standards.
8. Select or develop means for achieving the necessary emission reduction.
9. Decide the date after which each source type should be controlled.
10. Set emission standards.
11. Enforce emission standards.
12. Continue monitoring sources and the ambient atmosphere to ensure that adequate air quality is being achieved.

The Federal government is taking air management step 1. Expressed simply, the air quality criteria documents that are being issued list the observed effects of various pollutant concentrations and exposure durations on people, animals, plants, and property. The States and air quality control regions are basically responsible for taking the rest of the air management steps.

Air management step 2 involves deciding which air pollution effects to prevent. As an example that has been used several times, let us assume that an air management board wishes to achieve air quality that will not harm or offend man, animals, or plants.[2,3] Specifically, this would involve air quality such that:

1. The health of even sensitive or susceptible segments of the population would not be adversely affected.
2. Pollutant concentrations would not cause annoyances, such as unpleasant tastes or odors.

3. Damage to animals, ornamental plants, forests, and agricultural crops would not occur.
4. Visibility would not be significantly reduced.
5. Corrosion of metals and other materials would be minimal, but not completely eliminated.
6. Soiling, deterioration, or fading of fabrics would be minimal, but not completely eliminated.

Proceeding from Air Quality Criteria to Air Quality Standards

Air management step 3 involves selecting air quality standards that will prevent various unwanted effects. As a first example, let us consider the prevention of odors.

Odors

Dust and odors are the two chief air pollution complaints in Cincinnati, Ohio, according to Charles Gruber, former Chief Air Pollution Control Officer. Hemeon, in an article[4] titled "Sensory Air Pollution," notes that "Most of the objectionable effects of air pollution and the ones of dominant importance are those that assail the senses: those effects perceived as visible atmospheric haze and smoke, surface soiling due to smoke and dust, malodors of infinite variety, and irritation of the eyes." "Managing the Environment,"[2] a report from the U. S. House of Representatives, strikes the same theme: "The quality of the environment is not a human health issue, per se. It is more a matter of the unacceptability, at face value, of offensive odors, discolored water, low visibility, eye irritation, littered landscapes, and nuisance soiling." The point has also been made that Americans have been striving for better cars, refrigerators, television sets, and homes and that the American public is more than ready to strive for air of higher quality.

The subjective odor intensity detected by the human nose is not directly proportional to pollutant concentration. Rather, it is more nearly proportional to the logarithm of concentration (Weber-Fechner Law) as shown in Figure 1.[5]

$$\text{Odor intensity} = 1 + \log (\text{conc.})^{0.5} \qquad (1)$$

Thus, when the odor intensity is faint the concentration is 100 times that when the odor intensity is very faint; when the odor is easily noticeable the concentration is 100 times that for faint.

If the odor intensity of ethyl mercaptan in an area were easily noticeable, then 99 percent reduction in emissions would be needed to lower the intensity to faint (top axis of Figure 1), 99.99 percent control would be needed to lower it to very faint, and 99.9999 percent control would be needed to lower it to no odor. The human response to odors indicated in Figure 1 shows why odorous processes, such as rendering or kraft paper manufacturing,[6] require very efficient treatment of *all* emissions. Fortunately, efficient treatment is attainable by direct combustion. The rules and regulations of the

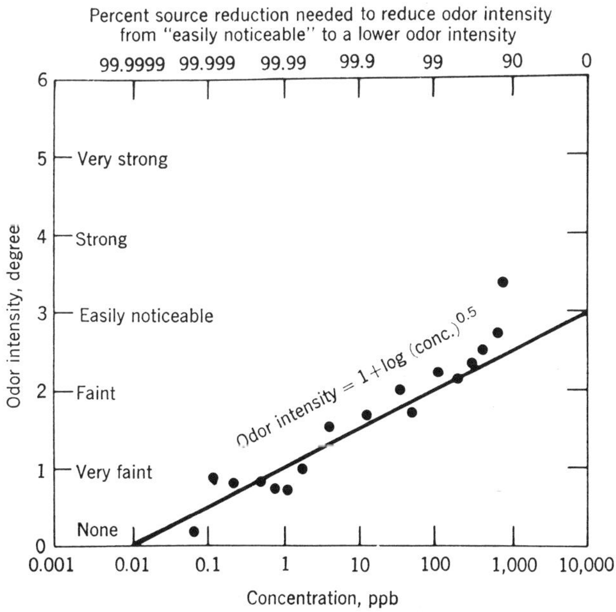

Figure 1. Odor intensity at various concentrations of ethyl mercaptan.

Los Angeles County Air Pollution Control District when implemented, for instance, give effective means for achieving odor control and could readily be used in other counties, regions, and States.

The Weber-Fechner Law states that the human senses register intensity as being proportional to the logarithm of the strength of the stimulus, regardless of which of the five stimuli (light, sound, odor, taste, and touch) are involved. The Weber-Fechner Law quite possibly also holds for man's subjective feeling about the dirtiness of the atmosphere. A substantial reduction in the concentration of small, visibility-obscuring particles might be necessary before citizens would feel that the air was much clearer.

Not only is the sensation of odor intensity proportional to the logarithm of concentration, but the difference among individuals in their ability to smell an odor also seems to follow a logarithmic function, and thus plots as a straight line on log-probability paper (See Figure 2).[7] The most sensitive of 33 panelists smelled hydrogen sulfide at about one part of hydrogen sulfide per billion parts of air (1 ppb). The least sensitive did not sense hydrogen sulfide until the concentration was increased tenfold to 10 ppb. To prevent citizen complaints from hydrogen sulfide, concentrations would thus probably need to be kept below 1 ppb.

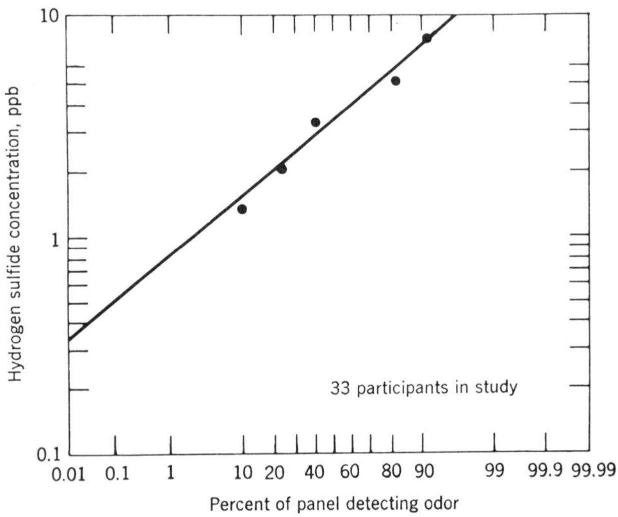

Figure 2. Percent of panel detecting odor at various hydrogen sulfide concentrations.

Another important aspect of odor is that the concentration need not remain above the odor threshold for long before the odor can be sensed. Odor can be sensed in a single, 1-sec-duration whiff. Just how do such short-duration concentrations relate to longer-term averages in the ambient atmosphere? Information on sulfur dioxide, another odorous pollutant, can be considered as an example.

Concentrations of sulfur dioxide and six other gaseous pollutants have been measured continuously for up to 7 years at Continuous Air Monitoring Program (CAMP) sites near the centers of Chicago, Cincinnati, Denver, Los Angeles, Philadelphia, St. Louis, San Francisco, and Washington, D. C. These concentration data have been analyzed as a function of averaging time and frequency and have been tabulated (Tables 1 through 7) for four different types of cities: a large midwest city, Chicago; a city with intense photochemical pollution, Los Angeles; a large eastern industrial city, Philadelphia; and a large non-industrial city, Washington. (Tables including the four other cities mentioned above are available upon request to the author.)

Analysis of aerometric data for carbon monoxide, hydrocarbons, nitric oxide, nitrogen dioxide, nitrogen oxides, oxidant, and sulfur dioxide indicates that concentration can be described by a mathematical model[8] with the following two characteristics:

1. Concentrations are approximately log normally distributed for all pollutants in all cities for all averaging times.
2. The median concentration (50 percentile) for all averaging times is proportional to averaging time to an exponent.

TABLE 1. CARBON MONOXIDE CONCENTRATION (PPM) AT CAMP SITES, BY AVERAGING TIME AND FREQUENCY, 1962-1968

CITY AND AVERAGING TIME	1962-1968 ANNL MAX	GEO MEAN	SGD	MAXIMA MAX	MAXIMA MIN	MAXIMUM FOR YEAR 62	63	64	65	66	67	68	PERCT. DATA AVAIL	PERCENT OF TIME CONC. IS EXCEEDED 0.01	0.1	1	10	30	50	70	90
CHICAGO																					
5 MIN	81	12.1	1.54	78	43	43	50	64	61	78	63	57	55	56	44	33	21	15	10	6	3
1 HR	53	12.3	1.46	59	28	28	36	46	44	59	56	40	56	51	40	30	21	15	10	7	3
8 HR	38	12.5	1.40	44	20	20	22	35	37	44	33	33	57		34	27	20	14	10	7	4
1 DAY	32	12.6	1.36	33	16	17	19	27	32	33	23	16	58			25	20	14	10	7	5
1 MONTH	18	13.0	1.22	21	7	7	10	17	21	18	11	7	62				18	13	10	7	6
1 YEAR	13	13.2	1.00	17	6			12	17	13		6	57						12		
LOS ANGELES																					
5 MIN	72	9.5	1.58	81	70	81	70	75					79	50	36	25	16	12	10	9	7
1 HR	46	9.7	1.50	47	35	43	35	47					79	43	34	25	16	12	10	9	7
8 HR	32	9.9	1.43	28	27	28	28	27					81		27	21	15	12	10	9	8
1 DAY	26	10.0	1.39	23	22	22	22	23					82			19	15	12	11	9	8
1 MONTH	15	10.3	1.24	14	13	14	13	14					89				13	12	11	9	9
1 YEAR	11	10.6	1.00	11	9	11	9	11					100						11		
PHILADELPHIA																					
5 MIN	83	6.7	1.77	67	43	51	52	43	67	47	44	49	58	47	35	22	13	9	7	5	3
1 HR	48	6.9	1.66	54	35	47	47	37	54	43	42	35	58	45	33	22	13	9	7	5	3
8 HR	31	7.1	1.56	36	23	36	35	27	26	34	23	26	59		27	20	12	9	7	5	3
1 DAY	24	7.2	1.50	25	14	23	25	21	19	20	14	23	60			18	12	9	7	6	4
1 MONTH	12	7.6	1.30	14	9		14	13	11	10	9	11	62				10	8	7	6	5
1 YEAR	8	7.9	1.00	8	6			7	8	7	6	8	71						7		
WASHINGTON																					
5 MIN	87	3.4	2.10	59	29	30	44	29	49	47	37	59	71			27	8	5	4	3	1
1 HR	44	3.6	1.93	41	25	25	41	28	32	38	32	27	71	35	27	17	8	5	4	3	2
8 HR	25	3.7	1.78	34	17	19	34	18	17	22	23	21	72		22	15	8	5	4	3	2
1 DAY	18	3.8	1.70	23	11	14	23	13	11	15	15	14	73			13	8	5	4	3	2
1 MONTH	7	4.2	1.41	10	5	7	10	6	6	5	7	7	76				7	5	4	3	2
1 YEAR	4	4.4	1.00	7	3		7	6	4	3	5	3	86						4		

TABLE 2. HYDROCARBON CONCENTRATION (PPM) AT CAMP SITES, BY AVERAGING TIME AND FREQUENCY, 1962-1968

CITY AND AVERAGING TIME	1962-1968 ANNL MAX	1962-1968 GEO MEAN	1962-1968 SGD	1962-1968 MAXIMA MAX	1962-1968 MAXIMA MIN	MAXIMUM FOR YEAR 62	63	64	65	66	67	68	PERCT. DATA AVAIL	PERCENT OF TIME CONC. IS EXCEEDED 0.01	0.1	1	10	30	50	70	90
CHICAGO																					
5 MIN	20	2.3	1.62	20	14	17	19	20	17	14	18	17	78	15	10	7	4	3	3	2	2
1 HR	12	2.4	1.53	13	8	10	12	8	10	11	13	12	78	12	9	6	4	3	3	2	2
8 HR	8	2.5	1.46	8	6	8	8	6	6	7	7	7	80		7	6	4	3	3	3	2
1 DAY	7	2.5	1.41	5	5	5	5	5	5	5	5	5	81			5	4	3	3	3	2
1 MONTH	4	2.6	1.25	4	3	4	4	3	3	3	3	3	88				4	3	3	3	2
1 YEAR	3	2.6	1.00	3	2	3	3	3	2	2	3	2	100						3		
LOS ANGELES																					
5 MIN	68	4.3	1.87	90	90		90						85	54	30	20	10	6	4	2	0
1 HR	38	4.5	1.74	30	30		30						85	28	25	19	10	6	4	2	0
8 HR	23	4.7	1.63	21	21		21						86		19	15	10	6	4	2	0
1 DAY	18	4.7	1.56	17	17		17						87			13	9	6	4	2	0
1 MONTH	8	5.0	1.34	9	9		9						100				7	5	3	2	1
1 YEAR	5	5.2	1.00	4	4		4						100						3		
PHILADELPHIA																					
5 MIN	20	2.3	1.62	17	10	12	15	17	10	14	14	13	70	12	9	6	4	3	2	2	1
1 HR	12	2.4	1.53	13	9	9	12	11	9	11	13	11	71	11	9	6	4	3	2	2	1
8 HR	8	2.5	1.46	8	7	7	7	8	7	8	7	8	72		7	5	3	3	2	2	2
1 DAY	7	2.5	1.41	6	4	5	6	6	5	5	6	4	73			5	3	3	2	2	2
1 MONTH	4	2.6	1.25	3	2	3	3	2	2	2	2	2	75				3	3	2	2	2
1 YEAR	3	2.6	1.00	2	2		2	2	2	2	2	2	86						2		
WASHINGTON																					
5 MIN	28	2.2	1.77	20	12	15	20	17	12	14	15	18	77	15	12	7	3	3	2	2	1
1 HR	16	2.3	1.66	17	9	12	17	14	9	12	15	13	77	14	11	7	3	3	2	2	1
8 HR	10	2.4	1.56	13	7	8	13	10	7	8	11	8	78		9	6	3	3	2	2	2
1 DAY	8	2.4	1.50	8	4	7	8	7	4	6	5	5	79			5	3	3	2	2	2
1 MONTH	4	2.5	1.30	3	2	3	3	3	3	2	3	3	83				3	3	2	2	2
1 YEAR	3	2.6	1.00	3	2		2	3	2	2	2	2	86						2		

TABLE 3. NITRIC OXIDE CONCENTRATION (PPM) AT CAMP SITES, BY AVERAGING TIME AND FREQUENCY, 1962-1968

CITY AND AVERAGING TIME	1962-1968 ANNL MAX	1962-1968 GEO MEAN	1962-1968 SGD	MAXIMA MAX	MAXIMA MIN	MAXIMUM FOR YEAR 62	63	64	65	66	67	68	PERCT. DATA AVAIL	PERCENT OF TIME CONC. IS EXCEEDED 0.01	0.1	1	10	30	50	70	90
CHICAGO																					
5 MIN	1.45	0.077	1.94	0.97	0.63	0.66	0.63	0.97	0.67	0.74	0.69	0.66	79	0.67	0.52	0.33	0.18	0.11	0.08	0.05	0.02
1 HR	0.77	0.081	1.80	0.91	0.60	0.61	0.60	0.91	0.62	0.64	0.63	0.61	79	0.64	0.50	0.32	0.18	0.11	0.08	0.05	0.02
8 HR	0.46	0.084	1.68	0.49	0.34	0.41	0.39	0.49	0.34	0.42	0.41	0.40	83		0.40	0.27	0.17	0.11	0.08	0.06	0.03
1 DAY	0.35	0.086	1.61	0.36	0.21	0.36	0.35	0.35	0.28	0.34	0.26	0.21	86			0.23	0.15	0.11	0.09	0.07	0.04
1 MONTH	0.15	0.092	1.36	0.17	0.11	0.17	0.15	0.16	0.13	0.14	0.15	0.11	94				0.13	0.11	0.08	0.07	0.06
1 YEAR	0.10	0.095	1.00	0.10	0.07	0.10	0.10	0.10	0.10	0.10	0.08	0.07	100						0.10		
LOS ANGELES																					
5 MIN	3.84	0.045	2.75	2.11	0.97	2.11	0.97	1.32					72	1.23	0.85	0.54	0.23	0.08	0.03	0.01	0.00
1 HR	1.54	0.050	2.45	1.42	0.79	1.42	0.79	1.24					72	1.12	0.80	0.53	0.23	0.08	0.03	0.01	0.00
8 HR	0.72	0.055	2.20	0.60	0.49	0.60	0.49	0.60					78		0.55	0.41	0.21	0.09	0.04	0.02	0.01
1 DAY	0.49	0.058	2.06	0.46	0.38	0.46	0.38	0.44					81			0.36	0.19	0.10	0.05	0.03	0.01
1 MONTH	0.15	0.067	1.60	0.25	0.18	0.25	0.18	0.19					92				0.16	0.09	0.06	0.03	0.02
1 YEAR	0.08	0.075	1.00	0.08	0.07	0.08	0.08	0.07					100						0.07		
PHILADELPHIA																					
5 MIN	6.54	0.023	3.61	1.98	1.03	1.58	1.72	1.35	1.03	1.98	1.74	1.68	75	1.60	0.95	0.37	0.12	0.05	0.03	0.01	0.00
1 HR	2.12	0.028	3.12	1.87	0.89	1.57	1.51	1.15	0.89	1.87	1.49	1.44	76	1.49	0.93	0.36	0.12	0.05	0.03	0.01	0.00
8 HR	0.84	0.032	2.71	1.00	0.52	0.93	0.91	0.52	0.53	1.00	0.72	0.87	80		0.72	0.31	0.11	0.06	0.03	0.02	0.01
1 DAY	0.52	0.035	2.50	0.58	0.25	0.46	0.58	0.25	0.31	0.48	0.34	0.37	82			0.25	0.10	0.06	0.04	0.02	0.01
1 MONTH	0.12	0.044	1.81	0.13	0.07	0.08	0.08	0.09	0.07	0.13	0.10	0.12	87				0.09	0.06	0.05	0.03	0.02
1 YEAR	0.05	0.053	1.00	0.06	0.04	0.04	0.05	0.04	0.04	0.06	0.06	0.06	100						0.04		
WASHINGTON																					
5 MIN	4.64	0.013	3.78	1.26	0.68	0.68	0.73	1.03	0.74	1.15	1.26	0.89	79			0.42	0.10	0.03	0.01	0.01	0.00
1 HR	1.46	0.016	3.25	1.14	0.62	0.65	0.68	0.88	0.62	1.02	1.14	0.69	80	0.88	0.62	0.33	0.09	0.03	0.01	0.01	0.00
8 HR	0.56	0.019	2.81	0.55	0.39	0.39	0.51	0.54	0.39	0.55	0.55	0.43	84		0.48	0.27	0.09	0.04	0.02	0.01	0.00
1 DAY	0.34	0.021	2.59	0.41	0.21	0.22	0.26	0.29	0.21	0.41	0.36	0.31	86			0.20	0.08	0.04	0.02	0.01	0.01
1 MONTH	0.07	0.027	1.85	0.10	0.06	0.07	0.08	0.07	0.06	0.09	0.08	0.10	93				0.07	0.04	0.03	0.02	0.01
1 YEAR	0.03	0.033	1.00	0.04	0.03	0.03	0.04	0.03	0.03	0.04	0.04	0.04	100						0.03		

TABLE 4. NITROGEN DIOXIDE CONCENTRATION (PPM) AT CAMP SITES, BY AVERAGING TIME AND FREQUENCY, 1962-1968

CITY AND AVERAGING TIME	1962-1968 ANNL MAX	GEO MEAN	SGD	MAXIMA MAX	MAXIMA MIN	MAXIMUM FOR YEAR 62	63	64	65	66	67	68	PERCT. DATA AVAIL	PERCENT OF TIME CONC. IS EXCEEDED 0.01	0.1	1	10	30	50	70	90
CHICAGO																					
5 MIN	0.45	0.037	1.76	0.79	0.21	0.26	0.23	0.79	0.21	0.35	0.29	0.29	79	0.28	0.18	0.13	0.07	0.05	0.04	0.03	0.02
1 HR	0.26	0.039	1.65	0.47	0.18	0.22	0.21	0.47	0.18	0.31	0.25	0.18	80	0.25	0.18	0.12	0.07	0.05	0.04	0.03	0.02
8 HR	0.17	0.040	1.55	0.22	0.12	0.16	0.17	0.22	0.15	0.20	0.15	0.12	84		0.16	0.12	0.07	0.05	0.04	0.04	0.03
1 DAY	0.13	0.040	1.49	0.15	0.09	0.12	0.13	0.15	0.09	0.15	0.11	0.10	87			0.10	0.07	0.05	0.04	0.04	0.03
1 MONTH	0.07	0.042	1.30	0.08	0.05	0.06	0.06	0.07	0.05	0.08	0.07	0.06	93				0.06	0.05	0.04	0.04	0.03
1 YEAR	0.04	0.044	1.00	0.06	0.04	0.04	0.04	0.05	0.04	0.06	0.05	0.05	100						0.04		
LOS ANGELES																					
5 MIN	1.48	0.038	2.30	1.27	0.57		0.57	1.27					63	0.66	0.44	0.25	0.12	0.06	0.04	0.03	0.01
1 HR	0.69	0.041	2.09	0.68	0.52		0.52	0.68					63	0.55	0.40	0.24	0.11	0.06	0.04	0.03	0.01
9 HR	0.36	0.044	1.91	0.37	0.27		0.37	0.27					67		0.27	0.23	0.11	0.06	0.05	0.03	0.02
1 DAY	0.26	0.045	1.81	0.26	0.20		0.20	0.26					71			0.18	0.10	0.07	0.05	0.04	0.02
1 MONTH	0.09	0.050	1.47	0.10	0.07		0.10	0.07					79				0.09	0.05	0.04	0.04	0.03
1 YEAR	0.05	0.054	1.00	0.06	0.05		0.06	0.05					100						0.05		
PHILADELPHIA																					
5 MIN	0.52	0.028	1.94	0.44	0.22	0.24	0.36	0.37	0.24	0.29	0.44	0.22	79	0.25	0.18	0.11	0.07	0.04	0.03	0.03	0.02
1 HR	0.28	0.030	1.80	0.32	0.20	0.22	0.32	0.26	0.20	0.23	0.23	0.20	79	0.23	0.18	0.11	0.07	0.04	0.04	0.03	0.02
8 HR	0.17	0.031	1.67	0.19	0.10	0.14	0.19	0.14	0.10	0.15	0.13	0.12	84		0.13	0.10	0.06	0.05	0.04	0.03	0.02
1 DAY	0.13	0.031	1.60	0.14	0.07	0.10	0.14	0.10	0.07	0.10	0.10	0.09	86			0.08	0.06	0.05	0.04	0.03	0.02
1 MONTH	0.06	0.033	1.36	0.06	0.04	0.05	0.06	0.05	0.04	0.05	0.05	0.05	90				0.05	0.04	0.04	0.03	0.03
1 YEAR	0.04	0.035	1.00	0.04	0.04	0.04	0.04	0.04	0.04	0.04	0.04	0.04	100						0.04		
WASHINGTON																					
5 MIN	0.38	0.030	1.79	0.37	0.19	0.37	0.24	0.24	0.25	0.19	0.24	0.26	79	0.22	0.16	0.10	0.06	0.04	0.03	0.03	0.02
1 HR	0.22	0.031	1.67	0.30	0.18	0.30	0.22	0.23	0.23	0.18	0.21	0.24	79	0.21	0.15	0.10	0.06	0.04	0.03	0.03	0.02
8 HR	0.14	0.032	1.57	0.14	0.09	0.10	0.11	0.14	0.10	0.09	0.14	0.13	84		0.11	0.09	0.06	0.04	0.03	0.03	0.02
1 DAY	0.11	0.032	1.51	0.10	0.07	0.07	0.09	0.10	0.07	0.07	0.09	0.08	86			0.08	0.06	0.04	0.04	0.03	0.02
1 MONTH	0.05	0.034	1.31	0.05	0.04	0.04	0.04	0.04	0.04	0.04	0.05	0.05	95				0.05	0.04	0.04	0.03	0.03
1 YEAR	0.03	0.035	1.00	0.05	0.03	0.03	0.03	0.04	0.03	0.03	0.04	0.05	100						0.03		

TABLE 5. NITROGEN OXIDES CONCENTRATION (PPM) AT CAMP SITES, BY AVERAGING TIME AND FREQUENCY, 1962-1968

CITY AND AVERAGING TIME	1962-1968 ANNL MAX	GEO MEAN	SGD	MAXIMA MAX	MAXIMA MIN	MAXIMUM FOR YEAR 62	63	64	65	66	67	68	PERCT. DATA AVAIL	PERCENT OF TIME CONC. IS EXCEEDED 0.01	0.1	1	10	30	50	70	90
CHICAGO																					
5 MIN	1.47	0.127	1.74	1.12	0.72	0.78	0.72	1.12	0.74	0.86	0.81		73	0.81	0.62	0.40	0.24	0.17	0.13	0.10	0.06
1 HR	0.86	0.132	1.63	1.06	0.68	0.69	0.68	1.06	0.68	0.75	0.74		74	0.76	0.60	0.39	0.24	0.17	0.13	0.10	0.06
8 HR	0.56	0.135	1.54	0.62	0.41	0.54	0.51	0.62	0.41	0.52	0.46		78		0.48	0.34	0.22	0.17	0.13	0.11	0.08
1 DAY	0.44	0.137	1.49	0.50	0.31	0.42	0.43	0.50	0.34	0.44	0.31		80			0.31	0.20	0.16	0.14	0.12	0.09
1 MONTH	0.22	0.144	1.29	0.22	0.17	0.21	0.18	0.22	0.17	0.19	0.19		88				0.17	0.15	0.14	0.13	0.11
1 YEAR	0.15	0.149	1.00	0.16	0.13	0.14	0.14	0.15	0.14	0.16	0.13		100						0.14		
LOS ANGELES																					
5 MIN	3.69	0.088	2.34	1.49	1.12		1.12	1.49					61	1.28	0.97	0.67	0.31	0.14	0.08	0.05	0.02
1 HR	1.68	0.095	2.12	1.39	1.07		1.07	1.39					61	1.09	0.97	0.67	0.31	0.14	0.08	0.05	0.02
8 HR	0.88	0.101	1.94	0.78	0.71		0.71	0.78					64		0.71	0.53	0.29	0.14	0.09	0.05	0.03
1 DAY	0.63	0.105	1.84	0.55	0.54		0.54	0.55					68			0.48	0.26	0.15	0.09	0.06	0.04
1 MONTH	0.22	0.116	1.48	0.28	0.23		0.28	0.23					79				0.23	0.11	0.09	0.08	0.05
1 YEAR	0.13	0.126	1.00	0.13	0.12		0.13	0.12					100						0.12		
PHILADELPHIA																					
5 MIN	4.80	0.056	2.75	2.16	1.14	1.81	1.93	1.50	1.14	2.16	1.87		70	1.81	1.04	0.42	0.17	0.10	0.07	0.05	0.03
1 HR	1.92	0.063	2.45	2.02	0.97	1.79	1.69	1.27	0.97	2.02	1.62		71	1.69	1.00	0.41	0.17	0.10	0.07	0.05	0.03
8 HR	0.90	0.069	2.20	1.11	0.45	1.06	1.06	0.45	0.60	1.11	0.79		75		0.80	0.36	0.16	0.10	0.07	0.05	0.03
1 DAY	0.61	0.072	2.06	0.70	0.28	0.56	0.70	0.28	0.38	0.58	0.42		76			0.29	0.15	0.10	0.08	0.06	0.04
1 MONTH	0.18	0.084	1.60	0.18	0.10	0.13	0.12	0.10	0.10	0.18	0.15		81				0.12	0.09	0.08	0.07	0.06
1 YEAR	0.09	0.094	1.00	0.10	0.08	0.08		0.08	0.08	0.10	0.10		83						0.08		
WASHINGTON																					
5 MIN	3.44	0.039	2.77	1.47	0.79	0.79	0.86	1.16	0.84	0.85	1.47		71	1.00	0.70	0.38	0.14	0.07	0.05	0.03	0.02
1 HR	1.37	0.044	2.46	1.30	0.68	0.74	0.76	1.00	0.71	0.68	1.30		71	0.97	0.71	0.38	0.14	0.07	0.05	0.04	0.02
8 HR	0.64	0.048	2.21	0.69	0.41	0.43	0.58	0.69	0.45	0.41	0.56		75		0.54	0.31	0.13	0.07	0.05	0.04	0.03
1 DAY	0.43	0.051	2.07	0.45	0.27	0.27	0.35	0.38	0.27	0.30	0.45		78			0.25	0.12	0.08	0.06	0.05	0.03
1 MONTH	0.13	0.059	1.60	0.13	0.10	0.11	0.13	0.12	0.10	0.12	0.11		86				0.10	0.08	0.06	0.05	0.04
1 YEAR	0.07	0.066	1.00	0.08	0.06	0.06	0.07	0.07	0.07	0.07	0.08		100						0.07		

TABLE 6. OXIDANT CONCENTRATION (PPM) AT CAMP SITES, BY AVERAGING TIME AND FREQUENCY, 1962-1968

CITY AND AVERAGING TIME	1962-1968 ANNL MAX	GEO MEAN	SGD	MAXIMA MAX	MAXIMA MIN	MAXIMUM FOR YEAR 62	63	64	65	66	67	68	PERCT. DATA AVAIL	PERCENT OF TIME CONC. IS EXCEEDED 0.01	0.1	1	10	30	50	70	90
LOS ANGELES																					
5 MIN	1.47	0.024	2.56	1.40	0.53	0.53	1.40	0.88					47	0.52	0.38	0.23	0.12	0.04	0.02	0.01	0.00
1 HR	0.62	0.026	2.30	0.49	0.41	0.49	0.49	0.41					69	0.47	0.34	0.21	0.10	0.04	0.02	0.01	0.00
8 HR	0.31	0.028	2.08	0.31	0.25	0.31	0.25	0.26					53		0.25	0.19	0.12	0.05	0.03	0.02	0.01
1 DAY	0.21	0.029	1.96	0.16	0.13	0.16	0.14	0.13					66			0.12	0.07	0.06	0.04	0.03	0.01
1 MONTH	0.07	0.033	1.55	0.07	0.06	0.06	0.06	0.07					28				0.06	0.06	0.06	0.04	0.02
1 YEAR	0.04	0.037	1.00										0						0.06		
PHILADELPHIA																					
5 MIN	0.93	0.018	2.45	0.60	0.12	0.12	0.13	0.25	0.25	0.60	0.29	0.31	72	0.51	0.24	0.13	0.06	0.03	0.02	0.01	0.00
1 HR	0.41	0.020	2.21	0.52	0.10	0.10	0.11	0.21	0.21	0.52	0.17	0.21	72	0.44	0.23	0.13	0.06	0.03	0.02	0.01	0.00
8 HR	0.21	0.021	2.01	0.35	0.08	0.08	0.08	0.14	0.14	0.35	0.13	0.13	78		0.18	0.11	0.05	0.03	0.02	0.02	0.01
1 DAY	0.15	0.022	1.89	0.18	0.05	0.05	0.05	0.09	0.09	0.18	0.08	0.08	79			0.08	0.05	0.03	0.02	0.02	0.01
1 MONTH	0.05	0.025	1.51	0.06	0.02	0.02	0.02	0.04	0.04	0.06	0.05	0.03	89				0.04	0.03	0.03	0.02	0.01
1 YEAR	0.03	0.027	1.00	0.03	0.01	0.01	0.01	0.02	0.03	0.03	0.03		100						0.02		
SAN FRANCISCO																					
5 MIN	0.54	0.012	2.35	0.42	0.29	0.32	0.29	0.42					65	0.26	0.15	0.07	0.04	0.02	0.02	0.01	0.00
1 HR	0.24	0.013	2.13	0.26	0.17	0.26	0.17	0.17					65	0.25	0.14	0.07	0.04	0.02	0.02	0.01	0.00
8 HR	0.13	0.014	1.94	0.16	0.10	0.16	0.10	0.11					69		0.11	0.06	0.03	0.02	0.02	0.01	0.01
1 DAY	0.09	0.015	1.84	0.09	0.05	0.09	0.08	0.05					69			0.06	0.03	0.02	0.02	0.01	0.01
1 MONTH	0.03	0.017	1.49	0.04	0.03	0.03	0.04	0.03					72				0.02	0.02	0.02	0.01	0.01
1 YEAR	0.02	0.018	1.00	0.02	0.02	0.02	0.02	0.02					100						0.02		
WASHINGTON																					
5 MIN	0.46	0.021	2.03	0.32	0.17	0.17	0.25	0.19	0.24	0.22	0.32	0.29	75	0.20	0.15	0.11	0.06	0.03	0.02	0.01	0.00
1 HR	0.24	0.022	1.87	0.26	0.13	0.13	0.22	0.18	0.21	0.16	0.26	0.25	76	0.20	0.15	0.11	0.06	0.03	0.02	0.01	0.00
8 HR	0.14	0.023	1.73	0.17	0.10	0.10	0.16	0.11	0.12	0.12	0.17	0.13	80		0.12	0.10	0.06	0.03	0.02	0.02	0.01
1 DAY	0.10	0.023	1.66	0.11	0.06	0.06	0.08	0.08	0.09	0.10	0.11	0.10	83			0.07	0.05	0.04	0.03	0.02	0.01
1 MONTH	0.04	0.025	1.39	0.04	0.03	0.03	0.03	0.04	0.04	0.04	0.04	0.04	89				0.04	0.03	0.03	0.02	0.01
1 YEAR	0.03	0.026	1.00	0.03	0.01	0.01	0.01	0.03	0.03	0.03	0.03	0.03	100						0.03		

Note—Sulfur dioxide interference caused low concentration readings for oxidant in 1962 and 1963. This fact shows especially for maximum concentrations for averaging times of 1 month and 1 year. Data for these 2 years were not used for the frequency distributions above, except in Los Angeles and San Francisco, where the interference was much less serious. Interference was controlled after 1963.

TABLE 7. SULFUR DIOXIDE CONCENTRATION (PPM) AT CAMP SITES, BY AVERAGING TIME AND FREQUENCY, 1962-1968

CITY AND AVERAGING TIME	ANNL MAX	1962-1968 GEO MEAN	SGD	MAXIMA MAX	MAXIMA MIN	MAXIMUM FOR YEAR 62	63	64	65	66	67	68	PERCT. DATA AVAIL	PERCENT OF TIME CONC. IS EXCEEDED 0.01	0.1	1	10	30	50	70	90
CHICAGO																					
5 MIN	3.27	0.104	2.19	1.94	1.11	1.13	1.94	1.62	1.59	1.11	1.62	1.33	83	1.31	1.01	0.67	0.32	0.16	0.08	0.03	0.01
1 HR	1.58	0.112	2.00	1.69	0.86	0.86	1.69	1.12	1.14	0.98	1.11	0.91	83	1.11	0.95	0.64	0.32	0.16	0.08	0.03	0.01
8 HR	0.86	0.118	1.84	1.02	0.45	0.45	0.87	1.02	0.74	0.63	0.80	0.63	84		0.74	0.56	0.31	0.16	0.09	0.04	0.01
1 DAY	0.63	0.121	1.75	0.79	0.36	0.36	0.71	0.79	0.55	0.49	0.65	0.51	86			0.49	0.30	0.16	0.09	0.05	0.02
1 MONTH	0.24	0.133	1.44	0.35	0.18	0.19	0.33	0.35	0.27	0.26	0.32	0.26	94				0.27	0.18	0.09	0.06	0.03
1 YEAR	0.14	0.142	1.00	0.18	0.09	0.10	0.14	0.18	0.13	0.09	0.12	0.12	100						0.12		
LOS ANGELES																					
5 MIN	0.47	0.013	2.28	0.68	0.24	0.24	0.51	0.68					59	0.38	0.15	0.08	0.04	0.02	0.01	0.01	0.00
1 HR	0.22	0.014	2.07	0.29	0.13	0.13	0.19	0.29					67	0.25	0.13	0.08	0.04	0.02	0.01	0.01	0.00
8 HR	0.12	0.015	1.90	0.13	0.08	0.08	0.09	0.13					71		0.10	0.07	0.04	0.02	0.02	0.01	0.00
1 DAY	0.09	0.015	1.80	0.10	0.06	0.06	0.07	0.10					71			0.06	0.04	0.02	0.02	0.01	0.00
1 MONTH	0.03	0.017	1.46	0.03	0.02	0.03	0.03	0.02					64				0.02	0.02	0.02	0.01	0.01
1 YEAR	0.02	0.018	1.00	0.02	0.02	0.02	0.02						67						0.02		
PHILADELPHIA																					
5 MIN	2.67	0.055	2.41	1.25	0.79	1.25	1.05	1.00	1.11	0.79	1.12	1.10	77	0.96	0.72	0.48	0.20	0.09	0.05	0.03	0.01
1 HR	1.19	0.050	2.18	1.03	0.66	1.03	0.85	0.84	0.94	0.66	0.77	0.88	78	0.85	0.67	0.45	0.20	0.09	0.05	0.03	0.01
8 HR	0.61	0.064	1.99	0.71	0.43	0.56	0.58	0.54	0.71	0.43	0.56	0.55	79		0.52	0.37	0.19	0.10	0.06	0.04	0.01
1 DAY	0.43	0.067	1.88	0.46	0.35	0.35	0.46	0.43	0.35	0.36	0.35	0.36	81			0.31	0.17	0.10	0.07	0.04	0.02
1 MONTH	0.15	0.075	1.50	0.16	0.12	0.13	0.12	0.15	0.13	0.13	0.13	0.16	88				0.12	0.10	0.08	0.07	0.04
1 YEAR	0.08	0.081	1.00	0.10	0.06	0.09	0.06	0.09	0.08	0.09	0.10	0.08	100						0.08		
WASHINGTON																					
5 MIN	1.14	0.040	2.14	0.87	0.42	0.42	0.56	0.87	0.44	0.47	0.44	0.53	70	0.49	0.35	0.21	0.10	0.06	0.03	0.02	0.00
1 HR	0.56	0.042	1.96	0.62	0.35	0.38	0.48	0.62	0.35	0.45	0.37	0.42	70	0.45	0.34	0.21	0.10	0.06	0.03	0.02	0.00
8 HR	0.31	0.045	1.81	0.35	0.22	0.23	0.35	0.32	0.27	0.34	0.22	0.22	71		0.26	0.18	0.10	0.06	0.04	0.02	0.01
1 DAY	0.23	0.046	1.72	0.25	0.15	0.18	0.25	0.22	0.20	0.25	0.15	0.18	72			0.16	0.10	0.06	0.04	0.02	0.01
1 MONTH	0.09	0.050	1.42	0.11	0.07	0.10	0.11	0.09	0.08	0.10	0.07	0.10	76				0.09	0.06	0.04	0.02	0.01
1 YEAR	0.05	0.053	1.00	0.05	0.04	0.05	0.05	0.04	0.05	0.04		0.04	86						0.04		

Two parameters will completely describe this mathematical model for a particular set of data. The two parameters that seem to work best for continuous concentration data are the 1-hr-averaging-time concentrations that are exceeded 0.1 and 30 percent of the time. These two points give a good fit to the data for the most polluted half of the air samples, and these are the samples of most interest in air management. These two points have been used in the model to calculate (1) the expected annual maximum concentration (expected to occur an average of once a year) listed in column 2 of Tables 1 through 7 and (2) the geometric means and standard geometric deviations listed in columns 3 and 4 of Tables 1 through 7.

The expected annual maximum concentrations listed in Tables 1 through 7 are suggested as relatively stable values to be used, as shown later, for calculating source reductions needed in a particular area. The annual maximum observed value often varies by a factor of 2 from year to year and is thus not recommended for such calculations. For instance, the maximum sulfur dioxide 1-hr-average concentration observed in Washington, D. C. during the 7-year period was 0.62 ppm, almost twice as high as the low value of 0.35 ppm observed in 1965. Columns 2 through 4 in Tables 1 through 7 are the calculated values mentioned above. All the rest of the values in Tables 1 through 7 are observed values.

Equations developed from the mathematical model allow calculation of annual maximum concentrations for various averaging times. These values for sulfur dioxide in Washington, D. C., are shown in column 2 of Table 7. The annual mean concentration is 0.05 ppm. The maximum day of 0.23 ppm is about 5 times the annual mean. The maximum hour of 0.56 ppm is about 10 times the annual mean. The calculated maximum second (not shown in table) of 6 ppm is about 10 times the maximum hour or 100 times the annual mean. The 1-hr-averaging-time standard geometric deviation of sulfur dioxide concentration in the CAMP cities varies from about 2 to 2.5 (Table 7, extended to include all 8 cities). When the standard geometric deviation is near 2, as it is for Washington, D. C., then the maximum second is expected to be about 10 times the maximum hour. When the deviation is near 2.5, then the maximum second is expected to be about 20 times the maximum hour.

A study of 53 odorant chemicals[9] indicates that all panelists recognized sulfur dioxide (as an oppressive odor) when the concentration was 0.47 ppm. The odor was not recognized at the next lowest concentration used, 0.047 ppm. Thus, the odor threshold apparently lies between 0.047 and 0.47 ppm and the sensory threshold (for odor or taste) is quite possibly near the 0.3 ppm mentioned in the air quality criteria document.[10] A 5.7 ppm reduction of the expected annual maximum 1-second-duration concentration of sulfur dioxide, therefore, would be needed to lower the 6 to 0.3 ppm. This 95 per cent reduction in the present concentration suggests

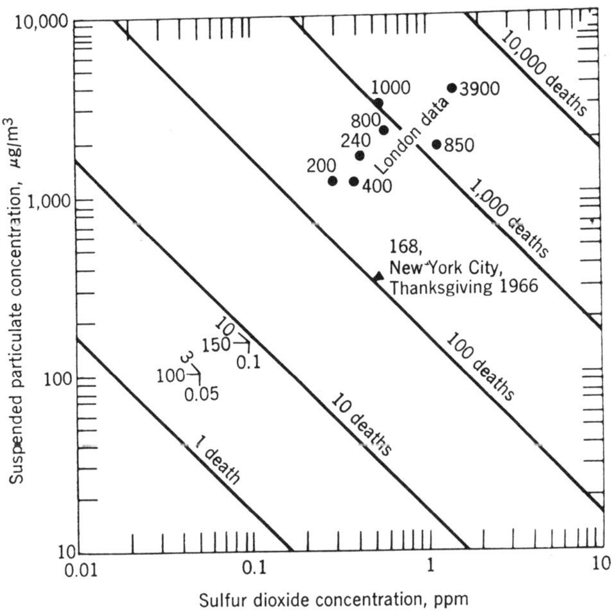

Figure 3. Number of excess deaths per episode expected in London or New York (population of about 10 million) when the maximum 24-hour-average concentrations of sulfur dioxide and suspended particulate reach various values.

that this degree of source reduction would be needed if the odor of sulfur dioxide were to be eliminated from Washington, D. C. Based on the sulfur content of present fuels, such control could be achieved either by treating stack gases or by restricting sulfur content of fuel to about 0.2 percent.

Death

Air quality criteria documents reference a multitude of research results, including studies of excess deaths attributed to air pollution episodes occurring in London and New York. One such study[11] discusses the excess deaths during 7 air pollution episodes in London. Each episode lasted several days. The study gives the maximum 24-hr-average concentrations of sulfur dioxide and suspended particulate experienced during each episode. This information could be used to estimate how many excess deaths might be prevented by controlling air pollution to various degrees. The data have been plotted (Figure 3) as excess deaths as a function of sulfur dioxide and particulate concentrations. The correlation and regression between deaths and pollutant concentration has been calculated here by taking logarithms of an equation of the form

$$D = ax^b \quad (2)$$

where D is the number of excess deaths, x is the pollutant concentration, and a and b are constants determined from the regression calculations. The resulting calculations indicate that the correlation of deaths against sulfur dioxide concentration is 0.87. The correlation of deaths against particulate concentration is 0.88. The exponent, b, is 1.5 for sulfur dioxide and 1.9 for particulate, suggesting a much stronger than linear relationship (exponent of 1) between

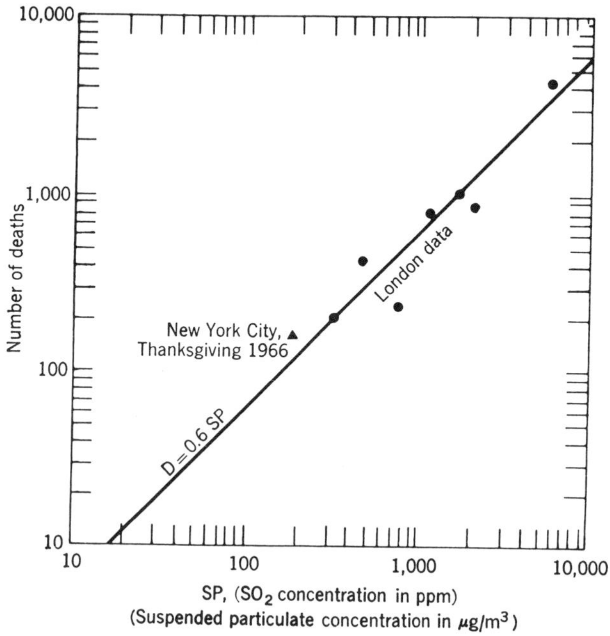

Figure 4. Number of deaths in London or New York air pollution episodes as a function of the product of sulfur dioxide and suspended particulate concentrations.

deaths and pollutant concentration. These results also suggest the possibility that deaths might be more closely related to the product of sulfur dioxide and particulate concentrations rather than to the concentration of either one of the pollutants alone.

Sulfur dioxide and particulate concentrations are not independent variables, but are highly correlated with each other. When atmospheric stagnation occurs, the concentrations of both pollutants rise. Over a period of years, however, particulate control in London has gradually lowered

38D. Thompson, R. B. and Roesner, G., Fluid Bed Roasting—Principles and Practice, in *Extractive Metallurgy of Copper, Nickel, and Cobalt,* Interscience, New York, 1961. pp. 3–32.
39D. Thoumsin, F. J. and Coussement, R., "Fluid-bed roasting reactions of copper and cobalt sulfide concentrates," *J. Metals* **16** (10), 831 (Oct 1964).
40D. Toguri, J. M., Themelis, N. J., and Jennings, P. H., "A review of recent studies on copper smelting," *Can. Met. Quart.* **3** (3), 197 (July 1964).
41D. Tsurumoto, T., "Copper smelting in the converter," *J. Metals* **13** (11), 820 (Nov 1961).
42D. Tsurumoto, T., Improvements on the Oxygen Smelting Process at Hitachi Smelter, in Anderson and Queneau, *Pyrometallurgical Processes in Nonferrous Metallurgy,* Gordon and Breach, New York, 1967. pp. 291–305.
43D. Worner, H. K., Continuous Smelting and Refining by the WORCRA Processes, *Advances in Extractive Metallurgy,* The Institution of Mining and Metallurgy, London, England, 1968. pp. 245–263.
44D. Zimmerley, S. R. (to Kennecott Copper Corp.) Reverberatory Smelting Method and Apparatus, U.S. Patent No. 3,102,806 (Sept. 3, 1963).

Hydrometallurgy

1E. Anon., "Ion exchange recovery of copper promising," *Chem. Eng. News* **45** (17), 62 (Apr. 17, 1967).
2E. Anon., "Pilot trials set for copper electrowinning," *Chem. Eng. News* **45** (26), 50 (June 19, 1967).
3E. Anon., "Bluebird Mine boosts output; New electrolytic copper plant to be built," *Eng. Mining J.* **168** (10), 118 (Oct 1967).
4E. Bhappu, R. B., "Chemical processing," *Mining Eng.* **20** (2), 99 (Feb 1968).
5E. Evans, D. J. I., Romanchuk, S., and Mackiw, V. N., "Treatment of copper-zinc concentrates by pressure hydrometallurgy," *Can. Mining Met. Bull.* **57** (628), 857 (Aug 1964).
6E. Evans, L. G. and Sheffer, H. W., "Dump and heap leaching practices in the Western United States," *Mining Congr. J.* **54** (9), 96 (Sept 1968).
7E. Forward, F. A. and Veltman, H., "Direct leaching zinc-sulfide concentrates by Sherritt Gordon," *J. Metals* **11** (12), 836 (Dec 1959).
8E. Habashi, F. and Torres-Acuña, N., "The anodic dissolution of copper (1) sulfide and the direct recovery of copper and elemental sulfur from white metal," *Trans. Met. Soc. AIME* **242** (5), 780 (May 1968).
9E. Mackiw, V. N., "Current trends in chemical metallurgy," *Can. J. Chem. Eng.* **46** (1), 3 (Feb. 1968).
10E. Mackiw, V. N., Benz, T. W., and Evans, D. J. I., "Recent developments in pressure hydrometallurgy," *Met. Reviews* **11** (109), 143 (Nov 1966).
11E. Renzoni, L. S., McQuire, R. C., and Barker, W. V., "Direct electrorefining of nickel matte," *J. Metals* **10** (6), 414 (June 1968).
12E. Rosenzweig, M. D., "New copper technology is winning the ore," *Chem. Eng.* **74** (25), 89 (Dec. 4, 1967).
13E. Spence, W. W. and Cook, W. R., "The Thompson Refinery," *Can. Mining Met. Bull.* **57** (631), 1181 (Nov 1964).
14E. Stephens, F. M., Jr. and Engle, L. F., "Hydrometallurgy—Today and tomorrow," *Mining Congr. J.* **54** (6), 37 (June 1968).
15E. Veltman, H., Pellegrini, S., and Mackiw, V. N., "Direct acid pressure leaching of chalcocite concentrate," *J. Metals* **19** (2), 21 (Feb. 1967).

16E. Vizsolyi, A., Veltman, H., and Forward, F. A., "Aqueous oxidation of galena in acid media under pressure," *Trans. Met. Soc. AIME* 227 (1), 215 (Feb 1963).
17E. Vizsolyi, A., Veltman, H., Warren, I. H., and Mackiw, V. N., "Copper and elemental sulfur from chalcopyrite by pressure leaching," *J. Metals* 19 (11), 52 (Nov. 1967).
18E. Warren, I. H., Vizsolyi, A., and Forward, F. A., "The pretreatment and leaching of chalcopyrite," *Can. Mining Met. Bull.* 61 (673), 637 (May 1968).

Emission Control Systems

1F. Anon., "Kennecott starts up huge H_2SO_4 plant," *Chem. Eng. News* 45 (13), 59 (Mar. 20, 1967).
2F. Anon., "Sulphur in its element," *Chemistry in Canada* 20 (4), 11 (Apr 1968).
3F. Anon., "New facility will produce iron-nickel pellets and elemental sulphur," *Eng. Mining J.* 169 (3), 168 (Mar 1968).
4F. Applebey, M. P., "The recovery of sulfur from smelter gases," *J. Soc. Chem. Ind.* 56, 139T (May 1937).
5F. Beals, G. C. and Cadle, J. J., "Copper and by-product sulfuric acid production by Palabora Mining Company," *J. Metals* 20 (7), 85 (July 1968).
6F. Burkhardt, D. B., "Increasing conversion efficiency," *Chem. Eng. Progr.* 64 (11), 66 (Nov. 1968).
7F. Carapella, S. C., Jr., Arsenic, Kirk-Othmer *Encyclopedia of Chemical Technology,* 2nd Ed., Vol. 2, 1963. pp. 711–717.
8F. Carter, B. M., Sulfuric Acid, Kirk-Othmer *Encyclopedia of Chemical Technology,* 1st Ed., Vol. 13, 1954. pp. 458–501.
9F. Connor, J. M., "Economics of sulfuric acid manufacture," *Chem. Eng. Progr.* 64 (11), 59 (Nov 1968).
10F. Dean, R. S., "Fixation of Sulfur from Smelter Smoke: Present Status of Sulfur Fixation and Plan of Investigations," U.S. Bureau of Mines, Rept. Invest. 3339 (May 1937).
11F. Donovan, J. R. and Stuber, P. J., "Sulfuric acid production from ore roaster gases," *J. Metals* 19 (11), 45 (Nov 1967).
12F. Donovan, J. R., and Stuber, P. J., "The technology and economics of interpass absorption sulfuric acid plants," Preprint No. 5E, A.I.Ch.E. 61st Ann. Mtg., Los Angeles, Calif. (Dec. 1–5, 1968).
13F. Duecker, W. W. and West, J. R. (Eds.), *The Manufacture of Sulfuric Acid,* ACS Monograph No. 144, Reinhold Publishing Co., New York, 1959.
14F. Fairlie, A. M., *Sulfuric Acid Manufacture,* ACS Monograph No. 69, Reinhold Publishing Co., New York, 1936.
15F. Farkas, M. D. and Dukes, R. R., "Multiple routes to sulfuric acid," *Chem. Eng. Progr.* 64 (11), 54 (Nov 1968).
16F. Fleming, E. P. and Fitt, T. C., "High purity sulfur from smelter gases," *Ind. Eng. Chem.* 42 (11), 2249 (Nov 1950).
17F. Fleming, E. P. and Fitt, T. C., "Liquid sulfur dioxide from waste smelter gases. Use of dimethylaniline as absorbent," *Ind. Eng. Chem.* 42 (11), 2253 (Nov 1950).
18F. Goar, B. G., "Today's sulfur recovery processes," *Hydrocarbon Processing* 47 (9), 248 (Sept 1968).
19F. Habashi, F., "The recovery of elemental sulfur from sulfide ores," State of Montana Bureau of Mines and Geology, Bulletin No. 51 (March 1966).
20F. Hensinger, C. E., Wakefield, R. E., and Claus, K. E., "Turning pollution gases into profits," *Eng. Mining J.* 169 (2), 131 (Feb 1968).

21F. Johnstone, H. F., "Recovery of sulfur dioxide from dilute gases," *Pulp Paper Mag. Can.* **53** (4), 105 (Mar 1952).
22F. Johnstone, H. F., Read, H. J., and Blankmeyer, H. C., "Recovery of sulfur dioxide from waste gases. Equilibrium vapor pressure over sulfite-bisulfite solutions," *Ind. Eng. Chem.* **30** (1), 101 (Jan 1938).
23F. Kelley, K. K., Contributions to the Data on Theoretical Metallurgy VII. The Thermodynamic Properties of Sulfur and Its Inorganic Compounds, U. S. Bureau of Mines Bulletin No. 406 (1937).
24F. Kirkpatrick, S. D., "Trail solves its sulfur problem," *Chem. Met. Eng.* **45** (9), 483 (Sept. 1938).
25F. Kronseder, J. G., "Cost of reducing sulfur dioxide emissions," *Chem. Eng. Progr.* **64** (11), 71 (Nov 1968).
26F. Lawler, W. C., Use of a Sulfur Dioxide Scrubbing System in Air Pollution Control, in F. S. Mallette, *Problems and Control of Air Pollution*, Reinhold Publishing Co., New York, 1955. pp. 222–224.
27F. Lehle, W. W., Processing of Waste Gases from Sulfuric Acid Plants, in Duecker and West, *The Manufacture of Sulfuric Acid*, Reinhold Publishing Co. New York, 1959. pp. 346–358.
28F. Lepsoe, R., "Chemistry of sulfur dioxide reduction. Thermodynamics," *Ind. Eng. Chem.* **30** (1), 92 (Jan 1938).
29F. Lepsoe, R., "Chemistry of sulfur dioxide reduction. Kinetics," *Ind. Eng. Chem.* **32** (7), 910 (July 1940).
30F. Lepsoe, R. and Kirkpatrick, W. S., "Sulfur dioxide recovery at Trail," *Trans. Can. Inst. Mining Met.* **40**, 399 (1937).
31F. McHaffie, I. R., Tyrer, D., and Imperial Chemical Industries, Ltd., Improvements in the Treatment of Metallic Sulfides to Recover Sulfur Therefrom. Brit. Pat. No. 425,364 (March 13, 1935).
32F. Moeller, W. and Winkler, K., "The double contact process for sulfuric acid production," *J. Air Poll. Control Assoc.* **18** (5), 324 (May 1968).
33F. Ontario Research Foundation, Removal of Sulphur Gases from Smelter Fumes, King's Printer, Toronto, Canada, 1949.
34F. Public Health Service and Manufacturing Chemists Association, Atmospheric Emissions from Sulfuric Acid Manufacturing Processes, Public Health Service Publication No. 999-AP-13 (1965).
35F. Queneau, P., Bracken, E. H., and Kelley, D., "High-grade iron ore at Copper Cliff, Ontario," *J. Metals* **10** (8), 527 (Aug 1958).
36F. Sawyer, F. G., Hader, R. N., Hernden, L. K., and Morningstar, E., "Sulfur from sour gases," *Ind. Eng. Chem.* **42** (10), 1938 (Oct 1950).
37F. Schackman, H., "Examples of new developments in ore processing resulting from chemical research," *Chem.-Ing.-Tech.* 26(2), 65 (Feb 1954).
38F. Scheidel, C. F., "Sulphur dioxide removal from tail gas by the Sulfacid Process," Preprint No. 6E, A.I.Ch.E. 61st Ann. Mtg., Los Angeles, Calif. (Dec. 1–5, 1968).
39F. Snowball, A. F. (Cominco Ltd., Trail, B.C.), Personal communication (Nov. 18, 1966).
40F. Srivastava, R. D., McMillan, A. F., and Harris, I. J., "The kinetics of oxidation of sodium sulphite," *Can. J. Chem. Eng.* **46** (3), 181 (June 1968).
41F. Stites, J. G., Jr., Horlacher, W. R., Jr., Bachofer, J. L. C., Jr., and Bartman, J. S., "The Catalytic-Oxidation System for Removing SO_2 from Flue Gas," Preprint No. 54 E, A.I.Ch.E. 61st Ann. Mtg., Los Angeles, Calif. (Dec. 1–5, 1968).

42F. Sulphur Patents Limited, Recovery of Sulphur from Smelter Gases, Kynoch Press, Birmingham, England (1936).
43F. Tarbutton, G., Driskell, J. C., Jones, T. M., Gray, F. J., and Smith, C. M., "Recovery of sulfur dioxide from flue gases," *Ind. Eng. Chem.* **49** (3), 392 (Mar 1957).
44F. Thompson R. and Butler, E. B., "History of sulfuric acid production from converter gas at the Kennecott Copper Corporation's Utah Smelter," *J. Metals* **20** (7), 82 (July 1968).
45F. Valdes, A. R., "New look at sulfur plants, Part 1: Design," *Hydrocarbon Processing*, **43** (3), 104 (Mar 1964).
46F. West, J. R. and Conroy, E. H., Process of Reducing Sulfur Dioxide to Elemental Sulfur, U. S. Pat. No. 3,199,955 (Aug. 10, 1965).

AUTHOR INDEX

KEY-WORD TITLE INDEX